AUTOMATIC CONTROL

a
SCIENTIFIC AMERICAN
book

SIMON AND SCHUSTER · NEW YORK

ALL RIGHTS RESERVED
INCLUDING THE RIGHT OF REPRODUCTION
IN WHOLE OR IN PART IN ANY FORM
© 1948, 1949, 1950, 1951, 1952, 1953, 1954 BY SCIENTIFIC
AMERICAN, INC.
© 1955 BY SCIENTIFIC AMERICAN, INC.
PUBLISHED BY SIMON AND SCHUSTER, INC.
ROCKEFELLER CENTER, 630 FIFTH AVENUE
NEW YORK 20, N.Y.

EIGHTH PRINTING
MANUFACTURED IN THE UNITED STATES OF AMERICA
PRINTED BY MURRAY PRINTING COMPANY
LIBRARY OF CONGRESS CATALOG CARD NUMBER: 55-12529

TABLE OF CONTENTS

INTRODUCTION   vii

1. FEEDBACK: THE PRINCIPLE OF CONTROL

    I.   SELF-REGULATION *by Ernest Nagel*   2
    II.   FEEDBACK *by Arnold Tustin*   10

With automatic control, as distinguished from mechanization in the familiar sense, we make machines regulate themselves. This is accomplished by the application of a simple, universal principle, called "feedback." It is feedback that makes life itself a self-regulating, i.e., an automatic, process. We shall see the principle applied not only to machines and automatic factories but to the administration of business and government. The steam engine took over the work of man's muscles; automatic control is the mechanization of brain.

2. THE SECOND INDUSTRIAL REVOLUTION

    I.   CONTROL SYSTEMS *by Gordon S. Brown and Donald P. Campbell*   26
    II.   AN AUTOMATIC CHEMICAL PLANT *by Eugene Ayres*   41
    III.   AN AUTOMATIC MACHINE TOOL *by William Pease*   53
    IV.   THE AUTOMATIC OFFICE *by Lawrence P. Lessing*   63
    V.   THE ECONOMIC IMPACT *by Wassily Leontief*   72

Though the automatic factory is still a thing of the future, large parts of our industrial system are already under automatic control. The robots do not merely save labor; they improve on human operators in many cases, and they manage some processes that lie hopelessly beyond man's ability to control. They are invading office as well as factory. So far, by higher wages and shorter hours, our economy has successfully absorbed the labor-saving impact of automatic control.

TABLE OF CONTENTS

3. INFORMATION: THE LANGUAGE OF CONTROL
    I. WHAT IS INFORMATION? *by Gilbert King*   83
    II. THE MATHEMATICS OF INFORMATION *by Warren Weaver*   97
    III. INFORMATION MACHINES *by Louis N. Ridenour*   111

Everything man has to say can be reduced to numbers. When this is done, information turns out to have the same mathematical qualities as entropy and is, therefore, subject to manipulation by the laws of thermodynamics. In numerical form, it can be processed quickly and accurately by mechanical computers or "information machines." These are the master machines of the future; they will control the automatic factory and make sense out of the enormous volumes of data involved in the administration of industry and government.

4. MACHINES AND MEN
    I. AN IMITATION OF LIFE *by W. Grey Walter*   123
    II. MAN VIEWED AS A MACHINE *by John G. Kemeny*   131

The "giant brains" now being put to use in science and industry are pale prototypes of those to come. Here are machines that behave as if they had a "will" of their own, that learn and improve on past performance and can reproduce their own kind. Such machines suggest deep questions about the human machine as well as immense new possibilities for advance of the art of automatic control.

BIBLIOGRAPHY   148

INTRODUCTION

Automatic control—or "automation," in the current jargon—is a major new movement in technology and an increasingly pervasive force in our social, political and economic life. It has been called a Second Industrial Revolution. Unfortunately, its arrival in public consciousness has been attended by as much misunderstanding as sensation.

There is going to be more and more automatic control in our lives. It is the means by which we will carry on big business and big government, production, finance, communications, trade and distribution in the complex and centrally organized civilization of our times. As citizens, we may hope to manage this revolution democratically and so to our advantage. We had better develop a clear understanding of what the term stands for and of the changes it implies in the way we make our livings and conduct our affairs.

This is the kind of interest on the reader's part to which this book is addressed. Its contents are the product of a unique collaboration between its scientist-authors and the editors of the magazine SCIENTIFIC AMERICAN *in which the chapters were first published during the past several years. Together, they provide a comprehensive picture of the unifying, essential principles of automatic control, of the degree to which the movement has already transformed our business and industrial system and an insight into the major lines of future development.*

One misunderstanding which this book clears up at the outset is that which confuses automatic control with mechanization in the familiar sense. The mechanization of labor has been going on for generations. Mechanical energy has long since displaced the biologically generated energy of man and beast in the day's work of our economy. The hewer of wood and the drawer of water has become the machine tender and the processor of paper. It is, in fact, the nervous systems, not the muscles, of men and women that our technology principally employs today. And it is the nervous system that is in process of replacement by automatic control.

Control is truly automatic when machines are made to regulate themselves. In every case, this is accomplished by the application of the same simple idea, which engineers call "feedback." Its more familiar illustration, a generation ago, would have been a steam engine and its flyball governor. Today, it is the household heating plant: the thermostat, a measuring device, responds to the output of heat from the furnace; it feeds back to the furnace a signal that regulates the input of fuel. This hookup of output to input brings about a radical change in the character of the system. The familiar chain of cause and effect—fuel, fire, heat—is closed into a loop of interdependent events. The heating system has become self-regulating. It is this idea, not a machine or a piece of hardware, that underlies automatic control.

As Ernest Nagel observes, feedback is an old idea. Life, a self-regulating process, is a feedback system; the principle is at work in the dynamic balance of nature and in the ebb and flow of human affairs. What is really new, Arnold Tustin shows, is our growing comprehension of the universality, virtuosity and power of the feedback principle.

Messrs. Brown and Campbell, in their survey, set straight another widespread misunderstanding. This is the notion that the automatic control revolution awaits the automatic factory. As yet, there is no such thing. But the robots, nonetheless, are here. In many industries, the robots already outnumber the workers and perform far more significant tasks.

The nearest we have yet come to a fully automatic factory is the petroleum refinery described by Eugene Ayres. It is operated by two men from a single, central control panel—two men, be-

INTRODUCTION

cause one might fall asleep or dead. But, as Mr. Ayres demonstrates, the achievement here is not the reduction of payroll. Modern refineries and chemical plants must be placed under automatic control because they are built to carry on processes that are too complex, too fast and too dangerous for control by human beings except through the mediation of robots.

The automatic factory as most people imagine it is previsioned in William Pease's description of an automatic machine tool. But here again the author shows that there are more important ends in view than depopulating the factory floor. Machines such as he describes would free our metal-working industries from the rigidities imposed by mechanization without automatic control, would give new flexibility and versatility to the small and medium-sized shop and vastly simplify the transfer of design from blueprint to finished product.

There is no doubt, however, that the advance of automatic control raises the serious issue of technological unemployment. Thanks to the vigilance of organized labor, public concern has thus far centered around the prospect of automatic factories. But Lawrence P. Lessing develops the unexpected conclusion that we should be much more immediately concerned with the arrival of the automatic office. White-collar functions, in the first place, employ nearly twice as many people as manufacturing. In the second place, it turns out, paper work is much more susceptible to automatic control than production. In government bureaus, banks, insurance companies, engineering departments, mail order houses, department stores, and accounting offices, machines of the most sophisticated design are taking over jobs by the hundreds. But even here payroll economics provides no exclusive motive. Our economic system has begun to outgrow the capacity of unaided man to control the information involved in its management.

Whether automatic control promises boon or distress is the difficult question faced by Wassily Leontief. From the record of the

INTRODUCTION

past, he infers a good prognosis. Mechanization has multiplied our output per man-hour by two in each generation for the past hundred years. Providing our capacity for social invention is equal to the task of securing a rational distribution of purchasing power, automatic control can continue to relieve men and women of unhealthy, unpleasant or unworthy work and give them increased material well-being and leisure.

With the question "What is Information?" Gilbert King introduces another central concept of automatic control. It has recently been realized that information of any kind, including the entire contents of the Library of Congress, can be expressed in numbers. Reduced to numbers, Warren Weaver shows, information assumes the same mathematical qualities as entropy and obeys the laws of thermodynamics. It is this new understanding of the nature of information that makes it possible for men to delegate so many thinking tasks to the mathematical machines described by Louis N. Ridenour.

In the last section of the book, we close what is perhaps the most sensitive and significant feedback loop in our culture. Advance in the technology of control, flowing from fundamental research, has now made it possible to build machines that lay profound challenges to fundamental research. W. Grey Walter's electrical tortoises demonstrate attributes of "free will" and immanent capacity to learn. John G. Kemeny describes the blueprint for a machine which, though it cannot make love, is capable of reproducing its kind. The thinking machines raise some deep questions about the nature of human thought itself.

THE EDITORS [*]

[*] Board of Editors: Gerard Piel (Publisher), Dennis Flanagan (Editor), Leon Svirsky (Managing Editor), George A. W. Boehm, Robert Emmett Ginna, Jean Le Corbeiller, James R. Newman, E. P. Rosenbaum, James Grunbaum (Art Director).

PART 1 **FEEDBACK: THE PRINCIPLE OF CONTROL**

I. **SELF-REGULATION** *by Ernest Nagel*

A leading contributor to the movement that is re-establishing the unity of science and philosophy, Ernest Nagel holds a chair in philosophy at Columbia University. He was born in Czechoslovakia in 1901 and came to the U. S. at the age of ten. After graduation from the College of the City of New York, he taught in the city's public schools and at his college while doing graduate work at Columbia. He took his doctorate in mathematical philosophy in 1931 and joined the Columbia faculty. Nagel is president of the Association for Symbolic Logic and author, with the late Morris Cohen, of *An Introduction to Logic and Scientific Method.*

II. **FEEDBACK** *by Arnold Tustin*

Arnold Tustin is professor of electrical engineering at the University of Birmingham, England, and a pioneer in the theory and practical technology of automatic control. In the latter field, he has been closely identified with application of advanced control methods to electrical transport, particularly to the London Underground system. In the realm of theory, he is the author of one of the standard epitomes with which engineers begin their acquaintance with the basic ideas and, more recently, of *Mechanisms of Economic Systems,* in which he develops the relevance of the concepts of control engineering to economics.

SELF-REGULATION
by Ernest Nagel

AUTOMATIC CONTROL is not a new thing in the world. Self-regulative mechanisms are an inherent feature of innumerable processes in nature, living and nonliving. Men have long recognized the existence of such mechanisms in living forms, although, to be sure, they have often mistaken automatic regulation for the operation of some conscious design or vital force. Even the deliberate construction of self-regulating machines is no innovation: the history of such devices goes back at least several hundred years.

Nevertheless, the preacher's weary cry that there is nothing new under the sun is at best a fragment of the truth. The general notion of automatic control may be ancient, but the formulation of its principles is a very recent achievement. And the systematic exploitation of these principles—their subtle theoretical elaboration and far-reaching practical application—must be credited to the twentieth century. When human intelligence is disciplined by the analytical methods of modern science, and fortified by modern material resources and techniques, it can transform almost beyond recognition the most familiar aspects of the physical and social scene. There is surely a profound difference between a primitive recognition that some mechanisms are self-regulative while others are not, and the invention of an analytic theory which not only accounts for the gross facts but guides the construction of new types of systems.

We now possess at least a first approximation to an adequate theory of automatic control, and we are at a point of history when the practical application of that theory begins to be conspicuous and widely felt. The future of automatic control, and the significance for human weal or woe of its extension to fresh areas of modern life, are still obscure. But if the future is not to take us completely by surprise, we need to survey, as the authors in this volume do, the principal content of automatic control theory, the problems that still face it and the role that automatic control is likely to play in our society.

The central ideas of the theory of self-regulative systems are simple and are explained with exemplary clarity in Mr. Tustin's essay which follows. Every operating system, from a pump to a primate, exhibits a characteristic pattern of behavior, and requires a supply of energy and a favorable environment for its continued operation. A system will cease to function when variations in its intake of energy or changes in its external and internal environment become too large. What distinguishes an automatically controlled system is that it possesses working components which maintain at least some of its typical processes despite such excessive variations. As need arises, these components employ a small part of the energy supplied to the system to augment or diminish the total volume of that energy, or in other ways to compensate for environmental changes. Even these elementary notions provide fruitful clues for understanding not only inanimate automatically controlled systems, but also organic bodies and their interrelations. There is no longer any sector of nature in which the occurrence of self-regulating systems can be regarded as a theme for oracular mystery-mongering.

However, some systems permit a greater degree of automatic control than others. A system's susceptibility to control depends on the complexity of its behavior pattern and on the range of variations under which it can maintain that pattern. Moreover, responses of automatic controls to changes affecting the operation of a system are in practice rarely instantaneous, and never absolutely accurate. An adequate science of automatic control must therefore develop comprehensive ways of discriminating and measuring variations in quality; it must learn how signals (or information) may be transmitted and relayed; it must be familiar with the conditions under which self-excitations and oscillations may occur, and it must devise mechanisms which will anticipate the probable course and sequence of events. Such a science will use and develop current theories of fundamental physico-chemical processes. It is dependent upon the elaborate logico-mathematical analyses of statistical aggregates, and upon an integration of specialized researches which until recently have seemed only remotely related. Our present theory of self-regulative systems has sprung from the soil of contemporary theoretical science. Its future is contingent upon the continued advance of basic research —in mathematics, physics, chemistry, physiology and the sciences of human behavior.

Automatic controls have been introduced into modern industry only in part because of the desire to offset rising labor costs. They are in fact not primarily an economy measure but a necessity, dictated by the nature of modern services and manufactured products and by the large demand for goods of uniformly high quality. Many articles in current use must be processed under

A generation ago, there would have been no need for a caption to explain the working of the flyball governor on the opposite page. Invented by James Watt in 1788 as a means for the automatic control of the output of a steam engine (it was, in fact, the crucial contribution made by Watt to the development of the steam engine), the spinning flyballs went out with the disappearance of the reciprocal steam engine. The flyball governor, however, has renewed significance today, as a symbol and a demonstration of the feedback principle that underlies all automatic control systems. Through the gears at the bottom of the vertical shaft in the picture, the governor was linked to the output shaft of the engine. The arm at the top of the shaft was linked, in turn, to the valve that controlled the input of steam to the engine. When the engine started up, the governor would begin to spin and the flyballs would be impelled outward from the governor shaft by centrifugal force. If the engine went too fast, the outward swing of the flyballs tended to close the input valve and slow the engine down; conversely, if the engine went too slow, the inward swing of the flyballs would tend to open the input valve. By this "feedback" of output to input, the engine was made to regulate itself.

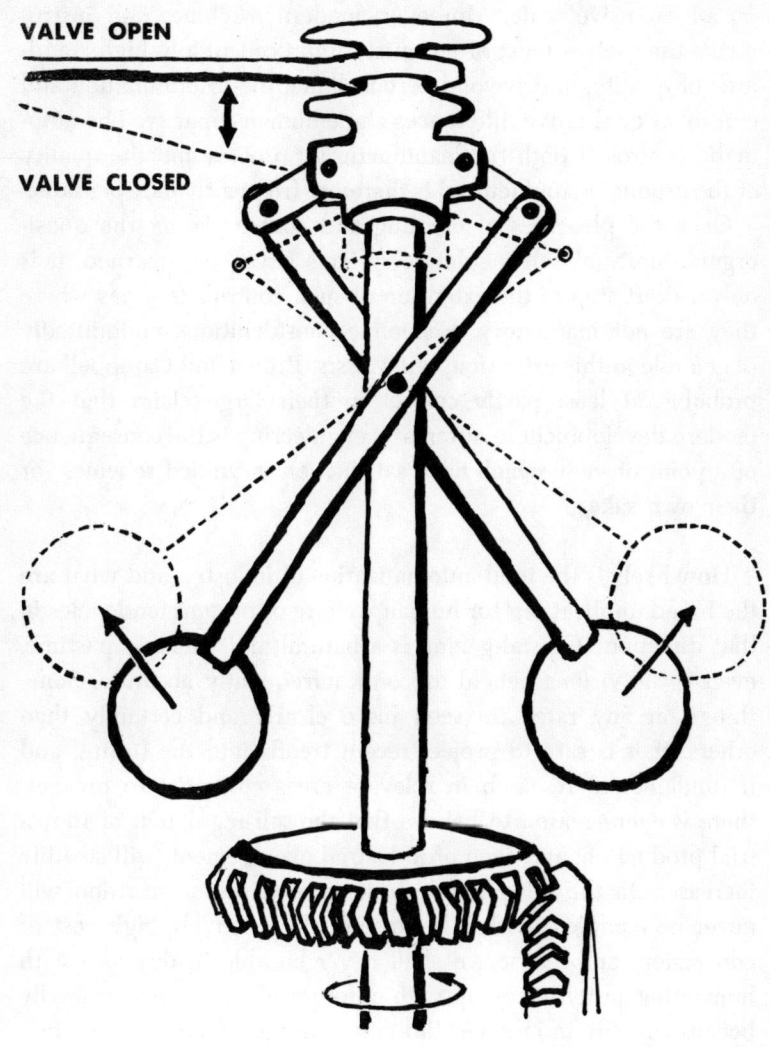

conditions of speed, temperature, pressure and chemical exchange which make human control impossible, or at least impracticable, on an extensive scale. Moreover, modern machines and instruments themselves must often satisfy unprecedentedly high standards of quality, and beyond certain limits the discrimination and control of qualitative differences elude human capacity. The automatic control of both the manufacturing process and the quality of the product manufactured is therefore frequently indispensable.

Once the pleasures of creating and contemplating the quasi-organic unity of self-regulative systems have been learned, it is only a short step to the extension of such controls to areas where they are not mandatory. Economic considerations undoubtedly play a role in this extension, but Messrs. Brown and Campbell are probably at least partly correct in their large claim that the modern development in automatic engineering is the consequence of a point of view which finds satisfaction in unified schemes for their own sake.

How likely is the total automatization of industry, and what are the broad implications for human welfare of present tendencies in that direction? Crystal-gazing is a natural and valuable pastime, even if the visions beheld are only infrequently accurate. Some things, at any rate, are seen more clearly and certainly than others. If it is safe to project recent trends into the future, and if fundamental research in relevant areas continues to prosper, there is every reason to believe that the self-regulation of industrial production, and even of industrial management, will steadily increase. On the other hand, in some areas automatization will never be complete—either because of the relatively high cost of conversion, or because we shall never be able to dispense with human ingenuity in coping with unforeseeable changes, or finally because of certain inherent limitations in the capacity of any machine which operates according to a closed system of rules. The dream of a productive system that entirely runs itself appears to be unrealizable.

SELF-REGULATION

Some consequences of large-scale automatic control in current technology are already evident, and are noted by several contributors to this volume. Industrial productivity has increased out of proportion to the increase in capital outlay. Many products are now of finer quality than they have ever been before. Working hours have been generally reduced, and much brutalizing drudgery has been eliminated. In addition, there are signs of a new type of professional man—the automatic control system engineer. There has been considerable conversion and retraining of unskilled labor. A slow refashioning of educational facilities, in content as well as in organization, in engineering schools as well as in the research divisions of universities and industries, is in progress. In the main these developments contribute to human welfare.

However, commentators on automatic control also see it as a potential source of social evil, and express fears—not altogether illegitimate—concerning its ultimate effect. There is first the fear that continued expansion in this direction will be accompanied by large-scale technological unemployment, and in consequence by acute economic distress and social upheaval. As Mr. Leontief points out in his chapter, the possibility of disastrous technological unemployment cannot be ruled out on purely theoretical grounds; special circumstances will determine whether or not it occurs. But, as he also notes, the brief history of automatic control in the U. S. suggests that serious unemployment is not its inevitable concomitant, at least in this country. The U. S. appears to be capable of adjusting itself to a major industrial reorganization without uprooting its basic patterns of living. Large-scale technological unemployment may be a more acute danger in other countries, but the problem is not insurmountable, and measures to circumvent or to mitigate it can be taken.

There is next the fear that an automatic technology will impoverish the quality of human life, robbing it of opportunities for individual creation, for pride of workmanship and for sensitive qualitative discrimination. This fear is often associated with a condemnation of "materialism" and with a demand for a return to the "spiritual" values of earlier civilizations. All the available evidence shows, however, that great cultural achievements are attained only by societies in which at least part of the population possesses considerable worldly substance. There is a good empirical basis for the belief that automatic control, by increasing the material well-being of a greater fraction of mankind, will release fresh energies for the cultivation and flowering of human excellence. At any rate, though material abundance undoubtedly is not a sufficient condition for the appearance of great works of the human spirit, neither is material penury; the vices of poverty are surely more ignoble than those of wealth. Moreover, there is no reason why liberation from the unimaginative drudgery which has been the lot of so many men throughout the ages should curtail opportunities for creative thought and for satisfaction in work well done. For example, the history of science exhibits a steady tendency to eliminate intellectual effort in the solution of individualized problems, by developing comprehensive formulas which can resolve by rote a whole class of them. To paraphrase Alfred North Whitehead, acts of thought, like a cavalry charge in battle, should be introduced only at the critical junction of affairs.

There has been no diminution in opportunities for creative scientific activity, for there are more things still to be discovered than are dreamt of in many a discouraged philosophy. And there is no ground for supposing that the course of events will be essentially different in other areas of human activity. Why should the wide adoption of automatic control and its associated quantitative methods induce a general insensitivity to qualitative distinctions? It is precisely measurement that makes evident the distinctions between qualities, and it is by measurement that man has frequently refined his discriminations and gained for them a wider acceptance. The apprehension that the growth of automatic controls will deprive us of all that gives zest and value to our lives appears in the main to be baseless.

There is finally the fear that an automatic technology will encourage the concentration of political power; that authoritarian controls will be established for all social institutions—in the interest of the smooth operation of industry and of society but to the ruin of democratic freedom. This forecast is given some substance by the recent history of several nations, but the dictatorships differ so greatly from the Western democracies in political traditions and social stratifications that the prediction has dubious validity for us. Nevertheless, one element in this grim conjecture requires attention. Whatever the future of automatic control, governmental regulation of social institutions is certain to increase—population growth alone will make further regulation imperative. It does not necessarily follow that liberal civilizations must therefore disappear. To argue that it does is to commit a form of the pathetic fallacy. Aristotle argued that political democracy was possible only in small societies such as the Greek city-states. If our present complex governmental regulations in such matters as sanitation, housing, transportation and education could have been foreseen by our ancestors, many of them would doubtless have concluded that such regulations are incompatible with any sense of personal freedom. It is easy to confound what is merely peculiar to a given society with the indispensable conditions for democratic life.

The crucial question is not whether control of social transactions will be further centralized. The crucial question is whether, despite such a movement, freedom of inquiry, freedom of communication and freedom to participate actively in decisions affecting our lives will be preserved and enlarged. It is good to be jealous of these rights; they are the substance of a liberal society. The probable expansion of automatic technology does raise serious problems concerning them. But it also provides fresh opportunities for the exercise of creative ingenuity and extraordinary wisdom in dealing with human affairs.

FEEDBACK
by Arnold Tustin

For hundreds of years a few examples of true automatic control systems have been known. A very early one was the arrangement on windmills of a device to keep their sails always facing into the wind. It consisted simply of a miniature windmill which could rotate the whole mill to face in any direction. The small mill's sails were at right angles to the main ones, and whenever the latter faced in the wrong direction, the wind caught the small sails and rotated the mill to the correct position. With steam power came other automatic mechanisms: the engine-governor, and then the steering servo-engine on ships, which operated the rudder in correspondence with movements of the helm. These devices, and a few others such as simple voltage regulators, constituted man's achievement in automatic control up to about twenty years ago.

In the past two decades necessity, in the form of increasingly acute problems arising in our ever more complex technology, has given birth to new families of such devices. Chemical plants needed regulators of temperature and flow; air warfare called for rapid and precise control of searchlights and anti-aircraft guns; radio required circuits which would give accurate amplification of signals.

Thus the modern science of automatic control has been fed by streams from many sources. At first, it now seems surprising to recall, no connection between these various developments was recognized. Yet all control and regulating systems depend on common principles. As soon as this was realized, progress became much more rapid. Today the design of controls for a modern boiler or a guided missile, for example, is based largely on principles first developed in the design of radio amplifiers.

FEEDBACK

Indeed, studies of the behavior of automatic control systems give us new insight into a wide variety of happenings in nature and in human affairs. The notions that engineers have evolved from these studies are useful aids in understanding how a man stands upright without toppling over, how the human heart beats, why our economic system suffers from slumps and booms, why the rabbit population in parts of Canada regularly fluctuates between scarcity and abundance.

The chief purpose of this essay is to make clear the common pattern that underlies all these and many other varied phenomena. This common pattern is the existence of feedback, or—to express the same thing rather more generally—interdependence.

We should not be able to live at all, still less to design complex control systems, if we did not recognize that there are regularities in the relationship between events—what we call "cause and effect." When the room is warmer, the thermometer on the wall reads higher. We do not expect to make the room warmer by pushing up the mercury in the thermometer. But now consider the case when the instrument on the wall is not a simple thermometer but a thermostat, contrived so that as its reading goes above a chosen setting, the fuel supply to the furnace is progressively reduced, and, conversely, as its reading falls below that setting, the fuel flow is increased. This is an example of a familiar control system. Not only does the reading of the thermometer depend on the warmth of the room, but the warmth of the room also depends on the reading of the thermometer. The two quantities are interdependent. Each is a cause, and each an effect, of the other. In such cases we have a closed chain or sequence—what engineers call a "closed loop" (see diagram on page 12).

In analyzing engineering and scientific problems it is very illuminating to sketch out first the scheme of dependence and see how the various quantities involved in the problem are determined by one another and by disturbances from outside the system. Such a diagram enables one to tell at a glance whether a system is an open or a closed one. This is an important distinction, because a

AUTOMATIC CONTROL

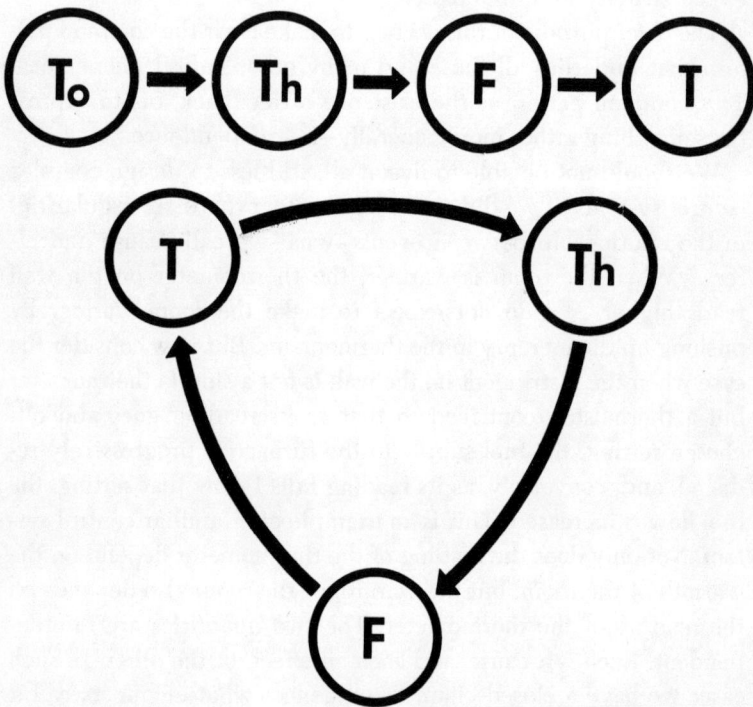

Feedback loop (at bottom) is here contrasted with open control sequence (top). In a hypothetical heating system, the outside temperature (To) might be employed to actuate the thermostat control (Th) on the furnace (F) to adjust room temperature (T). But there is no provision to determine whether the room temperature has attained the level desired. This self-regulating principle is uniquely provided by the feedback circuit. Here the variable which is to be controlled, the room temperature, itself actuates the thermostat. It thus controls the performance of the furnace.

closed system possesses several significant properties. Not only can it act as a regulator, but it is capable of various "self-excitatory" types of behavior—like a kitten chasing its own tail.

The now-popular name for this process is "feedback." In the case of the thermostat, the thermometer's information about the room temperature is fed back to open or close the valve, which in turn controls the temperature. Not all automatic control systems are of the closed-loop type. For example, one might put the thermometer outside in the open air, and connect it to work the fuel valve through a specially shaped cam, so that the outside temperature regulates the fuel flow. In this open-sequence system the room temperature has no effect; there is no feedback. The control compensates only that disturbance of room temperature caused by variation of the outdoor temperature. Such a system is not necessarily a bad or useless system; it might work very well under some circumstances. But it has two obvious shortcomings. Firstly, it is a "calibrated" system; that is to say, its correct working would require careful preliminary testing and special shaping of the cam to suit each particular application. Secondly, it could not deal with any but standard conditions. A day that was windy as well as cold would not get more fuel on that account.

The feedback type of control avoids these shortcomings. It goes directly to the quantity to be controlled, and it corrects indiscriminately for all kinds of disturbance. Nor does it require calibration for each special condition.

Feedback control, unlike open-sequence control, can never work without *some* error, for the error is depended upon to bring about the correction. The objective is to make the error as small as possible. This is subject to certain limitations, which we must now consider.

The principle of control by feedback is quite general. The quantities that it may control are of the most varied kinds, ranging from the frequency of a national electric-power grid to the degree of anesthesia of a patient under surgical operation. Control is exercised by negative feedback, which is to say that the information fed back is the amount of departure from the desired condition.

Any quantity may be subjected to control if three conditions are met. First, the required changes must be controllable by some physical means, a regulating organ. Second, the controlled quantity must be measurable, or at least comparable with some standard; in other words, there must be a measuring device. Third, both regulation and measurement must be rapid enough for the job in hand.

As an example, take one of the simplest and commonest of industrial requirements: to control the rate of flow of liquid along a pipe. As the regulating organ we can use a throttle valve, and as the measuring device, some form of flowmeter. A signal from the flowmeter, telling the actual rate of flow through the pipe, goes to the controller; there it is compared with a setting giving the required rate of flow. The amount and direction of "error," i.e., deviation from this setting, is then transmitted to the throttle valve as an operating signal to bring about adjustment in the required direction.

In flow-control systems the signals are usually in the form of variations in air pressure, by which the flowmeter measures the rate of flow of the liquid. The pressure is transmitted through a small-bore pipe to the controller, which is essentially a balance piston. The difference between this received pressure and the setting regulates the air pressure in another pipeline that goes to the regulating valve.

Signals of this kind are slow, and difficulties arise as the system becomes complex. When many controls are concentrated at a central point, as is often the case, the air-pipes that transmit the signals may have to be hundreds of feet long, and pressure changes at one end reach the other only after delays of some seconds. Meanwhile the error may have become large. The time-delay often creates another problem: overcorrection of the error, which causes the system to oscillate about the required value instead of settling down.

FEEDBACK

For further light on the principles involved in control systems let us consider the example of the automatic gun-director. In this problem a massive gun must be turned with great precision to angles indicated by a flypower pointer on a clock-dial some hundreds of feet away. When the pointer moves, the gun must turn correspondingly. The quantity to be controlled is the angle of the gun. The reference quantity is the angle of the clock-dial pointer. What is needed is a feedback loop which constantly compares the gun angle with the pointer angle and arranges matters so that if the gun angle is too small, the gun is driven forward, and if it is too large, the gun is driven back.

The key element in this case is some device which will detect the error of angular alignment between two shafts remote from each other, and which does not require more force than is available at the flypower transmitter shaft. There are several kinds of electrical elements that will serve such a purpose. The one usually selected is a pair of the miniature alternating-current machines known as selsyns. The two selsyns, connected respectively to the transmitter shaft and the gun, provide an electrical signal proportional to the error of alignment. The signal is amplified and fed to a generator which in turn feeds a motor that drives the gun.

This gives the main lines of a practicable scheme, but if a system were built as just described, it would fail. The gun's inertia would carry it past the position of correct alignment; the new error would then cause the controller to swing it back, and the gun would hunt back and forth without ever settling down.

This oscillatory behavior, maintained by "self-excitation," is one of the principal limitations of feedback control. It is the chief enemy of the control-system designer, and the key to progress has been the finding of various simple means to prevent oscillation. Since oscillation is a very general phenomenon, it is worth while to look at the mechanism in detail, for what we learn about oscillation in man-made control systems may suggest means of inhibiting oscillations of other kinds—such as economic booms and slumps, or periodic swarms of locusts.

Consider any case in which a quantity that we shall call the output depends on another quantity we shall call the input. If the input quantity oscillates in value, then the output quantity also will oscillate, not simultaneously or necessarily in the same way, but with the same frequency. Usually in physical systems the output oscillation lags behind the input. For example, if one is boiling water and turns the gas slowly up and down, the amount of steam increases and decreases the same number of times per minute, but the maximum amount of steam in each cycle must come rather later than the maximum application of heat, because of the time required for heating. If the first output quantity in turn affects some further quantity, the variation of this second quantity in the sequence will usually lag still more, and so on. The lag (as a proportion of one oscillation) also usually increases with frequency—the faster the input is varied, the farther behind the output falls.

Now suppose that in a feedback system some quantity in the closed loop is oscillating. This causes the successive quantities around the loop to oscillate also. But the loop comes around to the original quantity, and we have here the mechanism by which an oscillation may maintain itself. To see how this can happen, we must remember that with the feedback negative, the motion it causes would be opposite to the original motion, if it were not for the lags. It is only when the lags add up to just half a cycle that the feedback maintains the assumed motion. Thus any system with negative feedback will maintain a continuous oscillation when disturbed if (a) the time-delays in response at some frequency add up to half a period of oscillation, and (b) the feedback effect is sufficiently large at this frequency.

In a linear system, that is, roughly speaking, a system in which effects are directly proportional to causes, there are three possible results. If the feedback, at the frequency for which the lag is half a period, is equal in strength to the original oscillation, there will be a continuous steady oscillation which just sustains itself. If the feedback is greater than the oscillation at that frequency,

FEEDBACK

the oscillation builds up; if it is smaller, the oscillation will die away.

This situation is of critical importance for the designer of control systems. On the one hand, to make the control accurate, one must increase the feedback; on the other, such an increase may accentuate any small oscillation. The control breaks into an increasing oscillation and becomes useless.

To escape from the dilemma the designer can do several things. Firstly, he may minimize the time-lag by using electronic tubes or, at higher power levels, the new varieties of quick-response direct-current machines. By dividing the power amplification among a multiplicity of stages, these special generators have a smaller lag than conventional generators. The lag is by no means negligible, however.

Secondly, and this was a major advance in the development of control systems, the designer can use special elements that introduce a time-lead, anticipating the time-lag. Such devices, called phase-advancers, are often based on the properties of electric capacitors, because alternating current in a capacitor circuit leads the voltage applied to it.

Thirdly, the designer can introduce other feedbacks besides the main one, so designed as to reduce time-lag. Modern achievements in automatic control are based on the use of combinations of such devices to obtain both accuracy and stability.

So far we have been treating these systems as if they were entirely linear. A system is said to be linear when all effects are strictly proportional to causes. For example, the current through a resistor is proportional to the voltage applied to it; the resistor is therefore a linear element. The same does not apply to a rectifier or electronic tube. These are nonlinear elements.

None of the elements used in control systems gives proportional or linear dependence over all ranges. Even a resistor will burn out if the current is too high. Many elements, however, are linear over the range in which they are required to work. And when the range of variation is small enough, most elements will behave in an approximately linear fashion, simply because a very small bit of a curved graph does not differ significantly from a straight line.

We have seen that linear closed-sequence systems are delightfully simple to understand and—even more important—very easy to handle in exact mathematical terms. Because of this, most introductory accounts of control systems either brazenly or furtively assume that all such systems are linear. This gives the rather wrong impression that the principles so deduced may have little application to real, nonlinear systems. In practice, however, most of the characteristic behavior of control systems is affected only in detail by the nonlinear nature of the dependences. It is essential to be clear that nonlinear systems are not excluded from feedback control. Unless the departures from linearity are large or of special kinds, most of what has been said applies with minor changes to nonlinear systems.

Long before man existed, evolution hit upon the need for antioscillating features in feedback control and incorporated them in the body mechanisms of the animal world. Signals in the animal body are transmitted by trains of pulses along nerve fibers. When a sensory organ is stimulated, the stimulus will produce pulses at a greater rate if it is increasing than if it is decreasing. The maximum response, or output signal, occurs before the maximum of the stimulus. This is just the anticipatory type of effect (the time-lead) that is required for high-accuracy control. Physiologists now believe that the anticipatory response has evolved in the nervous system for, at least in part, the same reason that man wants it in his control mechanisms—to avoid overshooting and oscillation. Precisely what feature of the structure of the nerve mechanism gives this remarkable property is not yet fully understood.

Fascinating examples of the consequences of interdependence arise in the fluctuations of animal populations in a given territory. These interactions are sometimes extremely complicated. Charles Darwin invoked such a scheme to explain why there are more bumblebees near towns. His explanation was that near towns

FEEDBACK

there are more cats; this means fewer field mice, and field mice are the chief ravagers of bees' nests. Hence near towns bees enjoy more safety.

The interdependence of animal species sometimes produces a periodic oscillation. Just to show how this can happen, and leaving out complications that are always present in an actual situation, consider a territory inhabited by rabbits and lynxes, the rabbits being the chief food of the lynxes. When rabbits are abundant, the lynx population will increase. But as the lynxes become abundant, the rabbit population falls, because more rabbits are caught. Then as the rabbits diminish, the lynxes go hungry and decline. The result is a self-maintaining oscillation, sustained by negative feedback with a time-delay. This is not, of course, the complete picture of such phenomena as the well-known "fur cycle" of Canada, but it illustrates an important element in the mechanisms that cause it.

The periodic booms and slumps in economic activity stand out as a major example of oscillatory behavior due to feedback. In 1935 the economist John Maynard Keynes gave the first adequate and satisfying account of the essential mechanisms on which the general level of economic activity depends. Although Keynes did not use the terminology of control-system theory, his account fits precisely the same now-familiar pattern.

Keynes's starting point was the simple notion that the level of economic activity depends on the rate at which goods are bought. He took the essential further step of distinguishing two kinds of buying—of consumption goods and of capital goods. The latter is the same thing as the rate of investment. The money available to buy all these goods is not automatically provided by the wages and profits disbursed in making them, because normally some of this money is saved. The system would therefore run down and stop if it were not for the constant injection of extra demand in the form of new investment. Therefore the level of economic activity and employment depends on the rate of investment. This is the first dependence. The rate of investment itself, however, depends on the expectation of profit, and this in turn depends on the trend, present and expected, of economic activity. Thus not only does economic activity depend on the rate of investment, but the rate of investment depends on economic activity.

Modern theories of the business cycle aim to explain in detail the nature of these dependences and their characteristic nonlinearities. This clarification of the mechanisms at work immediately suggests many ways in which, by proper timing of investment expenditure, by more rational business forecasting, and so on, a stable level of optimal economic activity may be achieved in the near future. The day when it can unequivocally be said that slumps belong to the past will certainly be the beginning of a brighter chapter in human history.

The examples of feedback given here are merely a few selected to illustrate general principles. Many more will be described by other contributors to this volume. In this essay on "theory" I should like to touch on a further point: some ways in which the properties of automatic control systems or other complex feedback systems may be investigated in detail, and their performance perfected.

Purely mathematical methods are remarkably powerful when the system happens to be linear. Sets of linear differential equations are the happy hunting ground of mathematicians. They can turn the equations into a variety of equivalent forms, and gen-

Oscillation is inherent in all feedback systems. The drawing at top shows that when a regular oscillation is introduced into the input of a system (*lighter line*), it is followed somewhat later by a corresponding variation in the output of the system. The dotted rectangle indicates the lag that will prevail between equivalent phases of the input and the output curves. In the three drawings below, the input is assumed to be a feedback from the output. The first of the three shows a state of stable oscillation, which results when the feedback signal (*thinner line*) is opposite in phase to the disturbance of a system and calls for corrective action equal in amplitude. The oscillation is damped and may be made to disappear when, as in the next drawing, the feedback is less than the output. Unstable oscillation is caused by a feedback signal that induces corrective action greater than the error and thus amplifies the original disturbance.

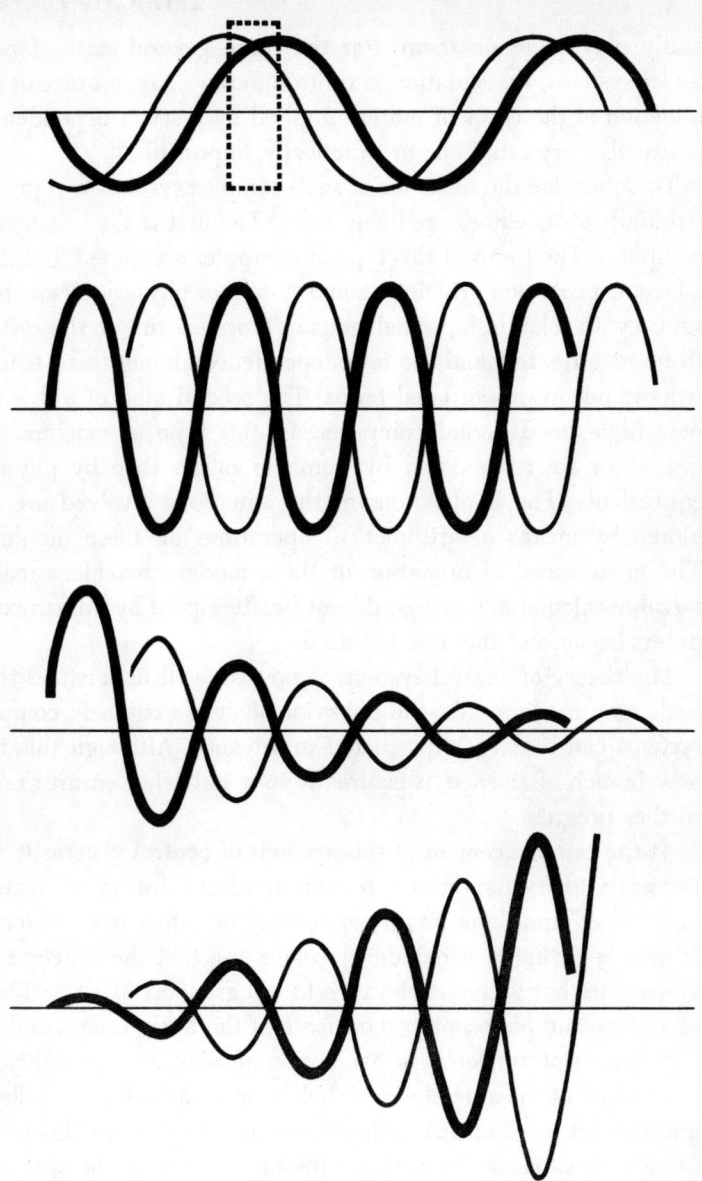

21

erally play tunes on them. For the more general class of non-linear systems, the situation is quite different. There exact determination of the types of motion implied by a set of dependences is usually very laborious or practically impossible.

To determine the behavior of such complex systems two principal kinds of machines are being used. The first is the "analogue" computer. The forms of this type of computer are varied, but they all share a common principle: some system of physical elements is set up with relationships analogous to those existing in the system to be investigated, and the interdependence among them is then worked out in proportional terms. The second kind of aid is the new high-speed digital computer. In this type of machine the quantities are represented by numbers rather than by physical equivalents. The implications of the equations involved are explored by means of arithmetical operations on these numbers. The great speed of operation of these modern machines makes possible calculations that could not be attempted by human computers because of the time required.

The theory of control systems is now so well understood that, with such modern aids, the behavior of even extremely complex systems can be largely predicted in advance. Although this is a new branch of science, it is already in a state that ensures rapid further progress.

At the commencement of this account of control systems it was necessary to assume that the human mind can distinguish "cause" and "effect" and describe the regularities of nature in these terms. It may be fitting to conclude by suggesting that the concepts reviewed are not without relevance to the grandest of all problems of science and philosophy: the nature of the human mind and the significance of our forms of perception of what we call reality.

In much of the animal world, behavior is controlled by reflexes and instinct-mechanisms in direct response to the stimulus of the immediate situation. In man and the higher animals the operation of what we are subjectively aware of as the "mind" provides a more flexible and effective control of behavior. It is not at present known whether these conscious phenomena involve potentialities of matter other than those we study in physics. They may well do so, and we must not beg this question in the absence of evidence.

Whatever the nature of the means or medium involved, the function of the central nervous system in the higher animals is clear. It is to provide a biologically more effective control of behavior under a combination of inner and environmental stimuli. An inner analogue or simulation of relevant aspects of the external world, which we are aware of as our idea of the environment, controls our responses, superseding mere instinct or reflex reaction. The world is still with us when we shut our eyes, and we use the "play of ideas" to predict the consequences of action. Thus our activity is adjusted more elaborately and advantageously to the circumstances in which we find ourselves.

This situation is strikingly similar in principle (though immensely more complex) to the introduction of a predictor in the control of a gun, for all predictors are essentially analogues of the external situation. The function of mind is to predict, and to adjust behavior accordingly. It operates like an analogue computor fed by sensory clues.

It is not surprising, therefore, that man sees the external world in terms of cause and effect. The distinction is largely subjective. "Cause" is what might conceivably be manipulated. "Effect" is what might conceivably be purposed.

Man is far from understanding himself, but it may turn out that his understanding of automatic control is one small further step toward that end.

PART 2 **THE SECOND INDUSTRIAL REVOLUTION**

I. **CONTROL SYSTEMS**
by Gordon S. Brown and Donald P. Campbell

In the minds of all students and practitioners of automatic control the names of Brown and Campbell are bracketed together as the authors of *Principles of Servomechanisms,* the basic handbook of their field. Gordon S. Brown was born in New South Wales, Australia, in 1907, got his undergraduate and graduate degrees at Massachusetts Institute of Technology and, after a brief period in professional service at home, returned to M.I.T. where he now heads the department of electrical engineering. Donald P. Campbell, born in New Jersey in 1918, went to Union College and then to M.I.T., where he is now associate professor of electrical engineering

II. **AN AUTOMATIC CHEMICAL PLANT**
by Eugene Ayres

A research chemist with wide industrial experience, Eugene Ayres is known outside his field as an authority on the energy resources and needs of the world economy. Born in South Carolina in 1891, he was educated at Swarthmore. From 1934 he was associated with the Gulf Research and Development Corporation and has recently retired to devote his time to research and writing on the energy problem. With Charles A. Scarlott, he is author of *Energy Sources.*

III. **AN AUTOMATIC MACHINE TOOL**
by William Pease

As associate professor of electrical engineering at M.I.T., William Pease was in charge of the development of the machine tool control system he describes here. Born in Rockland, Mass., in 1920, he was educated at M.I.T. Pease is now vice-president in charge of engineering of the Ultrasonics Corporation.

IV. THE AUTOMATIC OFFICE
by Lawrence P. Lessing

Lawrence P. Lessing went to work on a newspaper instead of going to college. He is an outstanding member of the small group of journalists who have made science reporting their specialty. At present, Lessing is working on a biography of the late Edwin H. Armstrong; he has been a member of the editorial boards of *Fortune* and SCIENTIFIC AMERICAN.

V. THE ECONOMIC IMPACT
by Wassily Leontief

Born in St. Petersburg, Russia, in 1905, Wassily Leontief received his undergraduate degree from the University of Leningrad and his Ph.D. from the University of Berlin. After serving as economic adviser to the Chinese government, he came to the U. S. in 1931, joining the faculty of Harvard University where he is professor of economics. He is author of *The Structure of the American Economy 1929–1939*, a work which demonstrated the practical utility of the Leontief "input-output" system, now being applied by many governments, including our own, to the analysis of national economies.

CONTROL SYSTEMS
by Gordon S. Brown and Donald P. Campbell

FEEDBACK CONTROL SYSTEMS have become our servants in many more ways than most of us realize. In the U. S. we now have tens of millions of them at work—in industry, in the military establishment, in offices and in the home. We are increasingly dependent upon this great army of robots. They minister to our comfort, protect our health and safety, relieve us of drudgery and operate difficult and dangerous enterprises which we would not dare undertake without them.

Everyone knows about the thermostat, which keeps radiators hot and the refrigerator cold. Most of the other controls are less familiar. A governor at the power station makes our electric clocks keep correct time. Electrical and electronic governors stabilize the performance of our radio and TV sets. A series of relays keeps the auto generator from overcharging the battery. Robots set the pitch of airplane propellers and trim the control surfaces to extract maximum performance from the engines and smooth the flight for the passengers. Process controllers supervise the manufacture of plastics, synthetic fibers, drugs, the whole range of products of the chemical industry. Intricate networks of instruments run our great petroleum refineries. Our communications system is one vast feedback circuit. Throughout industry the conversion of energy, from the process of combustion to the rotating of the shafts of heavy machinery, is conducted under automatic control. In sum, if the controls already operating in our economy were suddenly shut off, there would be chaos. The robots are here.

And this is only the beginning. The fully automatic factory is not yet here—but coming. The self-guiding missile is an inevitable

CONTROL SYSTEMS

prospect. One's imagination need not be restricted to industrial and military possibilities. We can look forward to feedback controls in homemaking, salesmanship, education, research, medicine and even contract-writing, designing and entertainment. If we can have feedback circuits in our radio and TV sets, why not an automatic program discriminator which could tell classical from popular music, if not Democratic from Republican orators? The thermostat suggests the possibility of a year-round program for the automatic heating and cooling of houses, responsive to the outdoor cycle of weather and season. Businessmen and administrators will think of ways to employ feedback circuits in such functions as inventory control and sales strategy. Physicians, who already use the feedback principle to control anesthesia and X-ray and diathermy treatments, will learn to extend the body's principle of homeostasis to control other aspects of therapy. In the sciences, we can look forward to probing the star-filled skies, the depths of the sea and the micro-dimensions of time and space with automatically controlled instruments.

A robot simulates the functions of a human being, and to understand how it works this is where we must start. Modern industry began with manual and semi-automatic controls. A human operator read the instruments and applied corrections to a process. He detected deviations of the actual performance from some desired standard and performed a corrective manipulation of valves, levers or rheostats. The human operator served as the feedback link, the error-detector, the controller and corrector. He decided what the reading was, what it meant, what should be done about correcting the process and whether or not the last correction was sufficient. This is the essence of control: measure, compare, correct and check the result.

As the tempo and complexity of technology rose, human operators began to fall behind: they could not keep up with the demands of the machine. It became necessary to give the instruments, which had merely measured and indicated, the function of control as well. Power amplification was added to them, to replace the service of human muscles. Mechanisms were devised to approximate some elementary abilities of the human brain—proportioning, anticipating and integrating. For many specific operations these devices closed the feedback control loop and made the human link unnecessary.

This liberation of technology from the limitations of the human system has already had great practical consequences—far greater than the labor-saving resulting from the reverse liberation of human beings from control jobs. It is revolutionizing the controlled processes themselves. Instead of merely hanging the new instruments on old plants, industry is redesigning the plants to take advantage of the opportunities offered by automatic control. Entirely new processes which deal with matter and energy in terms of quantum mechanics and probability, can now be engineered for control by feedback loops. This new approach is the province of a new and growing profession—feedback system engineering.

Servo-mechanisms, regulators and process controllers will thus play increasingly important roles. Simple controllers, which measure and control but one variable (e.g., temperature), are giving way to complex controllers which scan many variables, compute and govern the total plant performance. The robot specialists are being organized to function in teams. Combinations of automatic controllers and their respective processes function more and more as systems in the broadest sense—in energy conversion, transportation, communication, mechanized computation, the processing and synthesis of materials and the manipulation of machinery.

Let us consider three typical control circuits.

A red-hot ingot of steel as big as a ranch-house kitchen, moving ponderously down the ways from an oven, is suddenly caught by giant rollers. They whip it back and forth three or four times, pressing it down into a slab 100 feet long and six inches thick.

CONTROL SYSTEMS

The slab moves on to the trimming and cutting machines, and another ingot comes down the ways. Two men, sitting comfortably in leather chairs high above the heat and sparks, handle it all with levers. The rotating armatures of the electric motors that drive these mills are as big as an automobile; they deliver 5,000, sometimes 10,000, and next, perhaps, 25,000 horsepower. Yet these machines are lightning-fast. They zip the ingot from almost standstill to full speed in about one second; they stop it just as quickly.

The system "feels" the resistance of the ingot to the squeezing of the rolls and summons an exactly sufficient surge of current to keep the rolls turning at the speed demanded by the operator's hand on the lever. The circuit is closed at the control lever. Here the actual speed of the mill-rolls, continuously reported back by a tachometer on the spinning drive-shaft, is compared with the speed set by the operator's hand. Any error is transmitted to the drive as a tiny electrical signal and is amplified through a cascade of dynamos, each larger than the last, until it becomes the force needed to turn the rolls as directed. The human operator still controls other factors; he knows just how much to squeeze down the ingot at each pass through the rolls, and just how many times it must be passed to develop the metallurgical properties desired. But control engineers think that a feedback loop could take over these functions also. The equations for this "quality" loop will some day be written, bringing control of product quality and control of process together in a single system.

A radar dish tirelessly sweeps a beam of tiny electromagnetic pulses across the sea and sky. A target is discovered. The reflected pulses close a feedback circuit and freeze the radar on the target. Now the tracking of the dish puts computers to work, integrating the target position with weather and ballistic data. The solutions to these equations, fed through feedback circuits, start guns tracking the point in the sky where the trajectory of the shells and the course of the target will intersect. The final link in the system is a proximity fuse, which explodes a shell as it nears the target. This is diverse control, cutting across a half-dozen fields of engineering. Signals are translated with limitless virtuosity from discrete digital to continuous analogue statements, from electric and electronic impulses to mechanical and hydraulic action. But already these achievements in control technology are shaded by more recent developments in guided missiles.

The third illustration is a feedback loop which will control the ruling of diffraction gratings by a ruling engine. In a ruling engine a tiny diamond cuts a series of grooves in the aluminized surface of a small square of glass. The grooves, so close together that they are individually imperceptible to the human eye, must be exactly parallel, and they must be spaced accurately with a tolerance of less than a millionth of an inch. The problem of ruling a large grating (12 inches or more wide) with such accuracy has so far defied the machine and its human operators. But in the spectroscopic laboratory at the Massachusetts Institute of Technology a new automatically controlled engine is on the verge of achieving this objective. The measuring instrument in its feedback loop is an interferometer, which accurately measures distances as short as a wavelength of light. As the carriage on which the grating lies is shifted into position beneath the diamond for the ruling of each successive groove, the interferometer continuously measures its travel. If the carriage is more than one 50-millionth of an inch from the correct position, a signal is fed back to the motor that turns the screw that shifts the carriage, and it makes the necessary correction.

In an automatically controlled system, as in the human organism, the whole is far greater than its parts. The instruments, circuits, tubes and servo-mechanisms are but the hardware of the grand design. In the present state of the art there is a growing recognition among engineers and scientists that they cannot deal with control systems part by part, but must design each system as a unitary whole. Automatic control as we wish to speak of it here means the synthesis of product, process, plant and instruments. This implies designing the plant for control as much as designing controls for the plant. System engineering therefore calls for the pooled resources and efforts of professionals in many fields— mathematicians, scientists, engineers and administrators. It must integrate information and art from many branches of technology: mechanical, electrical, hydraulic, pneumatic, electronic, optical and chemical. Considering the specialization and complexity of each field, this is a formidable challenge. But the first generation of feedback system engineers—men who can grasp and synthesize the whole picture—is emerging. They have already left their mark on the chemical and electronics industries, in terms of better products and better co-ordination of plant operation.

Like many ideas which seem novel if we choose to ignore history, the system idea has a past. It began, in fact, with Watt's flyball-governor. This device betrays an inherent weakness of all feedback systems: a tendency to oscillate, or hunt, as Arnold Tustin explains in the preceding chapter. It was the mathematician who began to build the theory to bridge this difficulty. He showed how both oscillation and the damping of oscillation could be expressed in differential equations. By 1900 the theorems of Laplace and Fourier, the studies of Routh in analytical dynamics, the work of Kirchhoff in circuit analysis, the physical studies of Lord Kelvin and Heaviside and others had laid the foundations for a theory of control. But not until the 1920s did the exploitation of theory by practice really get under way.

During World War II great numbers of men were brought together from different fields to pool their abilities for the design of weapons and instruments. As a result, the specialists of engineering and science found themselves talking to one another for the first time in generations. Mechanical engineers exploited techniques of circuit theory borrowed from the communications engineers; aeronautical engineers extended the use of electrical concepts of measurement and of mathematical presentation; mathematicians working with engineers and experimental scientists discovered entirely unsuspected practical uses for forgotten theorems. The enforced collaboration soon focused attention on the essential principles that apply to all control systems. The general theory of control systems which now emerged was enriched in turn with the lore of experience from many different technologies. With the theoretical means at hand to write the equations for motors, amplifiers and hydraulic transmissions, it became possible to design control-system components with entirely new properties to meet predetermined needs. Moreover, these new parts could be used in many different types of control systems, and they could be manufactured in quantity.

It is one thing to design a system, and quite another to make it operate true to the theory. Its performance has to be tested and tuned. There were few precedents to guide the development of adequate testing methods. Earlier control systems had accepted some looseness in performance; their users were satisfied if equipment could be described as "fast" or "instantaneous," without more precise definition. In a radar feedback system, however, designers had to find ways to measure the dynamic, instant-by-instant behavior of each element to be sure that control signals would travel undistorted around the circuit. To accomplish this they developed various methods, borrowing ideas from half a dozen fields, particularly electrical and communications engineering.

The simplest way to test the behavior of the components in a circuit is to apply a sudden increase or decrease of voltage to the system. The shock produces different reactions in the various elements of the circuit, and measurement of the transient variations in voltage in the individual components provides important information about the performance of those parts and about the layout of the circuit as a whole. This so-called "step-function" test proved a valuable method of testing feedback circuits. Moreover,

CONTROL SYSTEMS

an analogous method can be used to test nonelectrical systems. A disturbing force or a hammer blow applied to the proper point sets up transient vibrations, corresponding to the voltage variations in a circuit. Plotting of these vibrations and of the time required for their decay can help the mechanical engineer to determine the mechanism's natural modes of vibration and the damping forces inherent in the system.

Another test for feedback circuits was based on one used in communications work. A radio or telephone message, transmitting music or a voice, is made up of a certain range of frequencies. In testing an amplifier, it is important to measure its performance in amplifying each of those frequencies. Communications engineers do so by applying a modified, distinguishable signal at each frequency and observing how the amplifier handles that frequency. Thus they can find out what will happen to a complex message in terms of what happens to the individual frequency components of the message. In the same way it is possible to test the behavior of equipment in a feedback circuit at various frequencies, and thereby to map the behavior pattern of the whole system. The frequency-response test (see diagram on page 35) opened the way to precise study of the dynamic behavior of such mechanisms as hydraulic transmissions and motor-generator sets. It defined the ability of these machines to transmit messages over wide ranges of operation—fast or slow, at low-power and high-power levels. Is it strange that high-power units such as hydraulic drives should be appraised in terms of their message-transmitting ability? Not at all, for in all feedback control systems the performance depends uniquely upon how the signals are acted upon by each and every mechanism operating in the closed path—individually and in unison. The practical bearing of this fact was illustrated vividly during the war: there were times when it was found that the entire performance of a $50,000 anti-aircraft gun-director was ruined by the faulty behavior of just one $50 torque motor in the system.

World War II was fought with servo-controlled turrets, servo-steered torpedoes, radar eyes and closed-loop missiles like the V-1, carrying a black box of automatic guidance. After the war the momentum of automatic control did not slacken but speeded up. The fever is still rising. We seem to be rushing forward both in practice and in the state of the art. Men excitedly speak of the robot factory, of the remotely controlled spaceship, of giant submarines that will respond to the touch of a little finger, of incredible new industrial processes. New industries are springing up every month to build heretofore unheard-of mechanisms. Older industries have shifted over to new lines of equipment. Researchers are probing new frontiers. Their work is not focused solely on iron mechanisms. We have taken a sudden new interest in the feedback mechanisms of the human system—chemical, physiological and psychological. Already we have the artificial feedback kidney and the mechanical heart.

It is not hard to view the race toward automatization with alarm. Can we realize all the dreams with a reasonable chance of avoiding economic chaos and without reducing the human nervous system to a state of uncontrolled oscillation?

The problems of system engineers grow steadily more difficult. They have encountered physical systems which do not respond to conventional mathematical treatment: for example, short-lived

Frequency response of two boilers with high (*left*) and low (*right*) heat-holding capacity is here contrasted. When the heat input to the two systems is varied regularly at the same frequency (*black curves in the center charts*), the temperature in each boiler (*gray curves in the center charts*) responds differently. The temperature in the high-capacity boiler varies with smaller amplitude and lags further behind the input frequency than that of the low-capacity system. A similar contrast between the two systems is developed by an abrupt increase in the heat input, as from the base line to the dotted line in the lower diagrams. In the high-capacity boiler, the temperature ascends slowly and oscillates gently above and below the set point before it levels off. The temperature curve in the low-capacity boiler shows a steeper increase, oscillates at higher frequency and damps out sooner. Response tests are made to determine performance of control systems.

LIQUID IN **LIQUID IN** **HEATING MEDIUM IN**

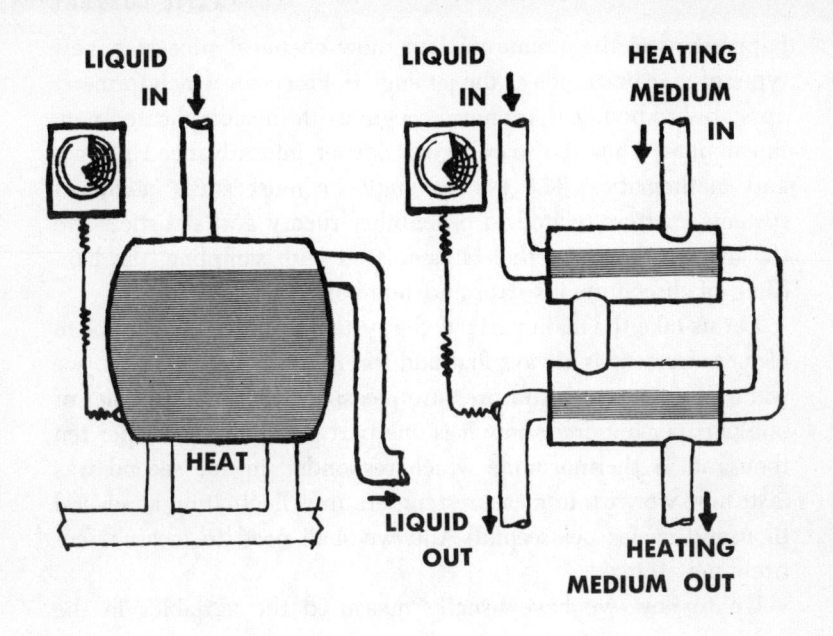

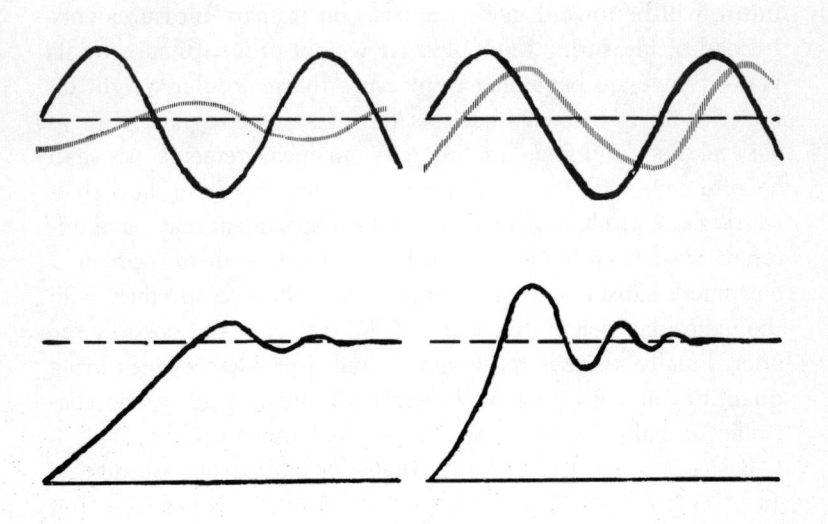

HEAT **LIQUID OUT** **HEATING MEDIUM OUT**

35

happenings in the atomic nucleus, new chemical processes, new types of machines such as the jet engine. Processes which formerly appeared to be linear are now recognized to be complicated nonlinear ones. Thus the search goes deeper into advanced physics and mathematics. The system engineer must study nonlinear systems, matters related to probability theory and statistics, and the new mathematics that is associated with sampling, the handling of discontinuous data and number theory.

Let us take the main parts of the system in order. The problem of measurement is always first and foremost. A few years ago we wanted temperature and pressure measurements to accuracies of one part per hundred; now it is one part per thousand or per ten thousand. A thermocouple which responded in one second was fast; now we wish to measure temperatures fluctuating at several thousand cycles per second! And we shall need to measure entirely new things.

Up to now we have usually measured the variables in the production process, such as flow, temperature and pressure, rather than the character of the product itself. The trend in the future will be toward more emphasis on primary measurements. Instead of measuring the molecular weight of a substance by its viscosity, we are beginning to measure the molecular weight directly, with instruments such as the mass spectrograph. But as we seek finer and finer resolution in our measurements, we shall have to cope with the interference of noise. Moreover, how shall we measure product quality? The sales department may have different ideas from the engineers. It may be difficult to correlate a customer's satisfaction and pleasure in a chemical product with the molecular weight, the shape of the molecule and possibly the odor. Finally, there is the tough technical problem of measuring quantities in a way that will enable a human or automatic controller to make decisions, e.g., as to which valve to turn.

Basically control is no better than the intelligence we give it, and this applies to human as well as to automatic control. But

CONTROL SYSTEMS

the engineers are not being caught napping. They have on their drafting boards designs for yet more discerning primary detectors, faster-responding instruments, plants capable of swifter and steadier production. And these engineers are conscious that quality of product is the ultimate objective, even though we may lack as yet the instruments with which to measure it fully.

At the next link in the control loop, the controller itself, equally fundamental changes are on the way. The typical industrial controller today is a simple analogue computer. It can deal only with one variable. The analogue device compares the measuring instrument's report on this variable with a set point fixed by the human operator, and generates a correcting signal proportional to the error. In the future, as Louis Ridenour explains (see page 111), the versatile digital computer will take over master control, perhaps supervising analogue devices assigned to specific jobs. The digital computer can handle many variables and provide continual solutions of complex groups of equations.

The controllers will have the problem of processing and translating the messages from the measuring instruments. For example, a message from an analogue device, say a voltage, must be translated into pulses to be handled by a digital computer, and the pulses in turn must be converted to another form of energy to jog a valve. Again, before the error signal can reach an appropriate valve or switch, the message may have to be sent long distances over pipelines, wires, radio links or even television channels. Single messages and multiple messages must be transmitted. And often they must be coded, among other reasons in order to avoid fouling one another. We may yet see two guided missiles tune in on each other's data systems and fall to gyrating around instead of closing in for a hit.

Redesign is also in store for the actuator elements, the muscles of the feedback loop. Many present control systems use pneumatic energy, especially where the fire hazard from electric sparking is great. An air-pressure signal comes from the controller to a big diaphragm, which huffs and puffs to move the valve-stem. In control systems of the future the valve-actuators will need more zip. New motors of all types, tiny in size, but with greater power, are required. Already there are miniature flow-valves, no thicker than a lead pencil, which can control hydraulic motors of one to five horsepower. By proper study of the heat-dissipation problem and by addition of cooling, such systems may become small enough to be carried in a coat pocket. At the other extreme, we shall need actuators of perhaps as much as 200,000 horsepower.

Feedback control-system engineering is a rapidly growing profession. The proper use and training of this new kind of talent demand a new type of thinking, both on the part of engineering management and of engineering schools. Every level of our industrial force, from the financial managers down to the maintenance workers, must be prepared to become acquainted with the benefits, operation and limitations of automatic control. Compromises must be made between costs and performance. Old methods of process design may be forced to yield to new ones, taking into account the limits of instruments rather than the cost and strength of steel tanks.

Control-system engineers will need a broad engineering education, an understanding of both the theory and practice of automatic control and an environment in which people are not afraid to experiment.

The training of such men offers educators a real challenge. Engineering schools will have to organize a new kind of program for them. The schools are usually organized in departments, such as mechanical and electrical, and within these divisions specialties like thermodynamics, electronics and circuit theory may be isolated areas of activity. A systems engineer cannot be trained by simply adding together the old specialties. What is wanted is not a jack-of-all-trades but a master of a new trade, and this will require a new synthesis of studies. It will call for advanced work in the fields of mathematics, physics, chemistry, measurements, communications and electronics, servo-mechanisms, energy conversion, thermodynamics and computational techniques. The control engineer will need to know the mathematics of differential equations, functions of a complex variable, statistics and nonlinear techniques, and to have a thorough grounding in modern physics and chemistry. He will also need to be familiar with computational aids, such as differential analyzers and computers.

Industrial management also must raise its thinking to the system level. Its technical staff, headed by a qualified system engineer, will have to work as a team; the chief electrical engineer cannot have exclusive control over all matters electrical, nor the mechanical engineer over all things mechanical.

Many schools are now offering short concentrated programs on feedback controls, computers, advanced mathematics, information theory and statistical communication theory for engineers in industry and research centers.

Most businessmen, understandably, have forgotten their calculus and advanced mathematics. They may nevertheless want to know what the system engineer means when he speaks of "process lags" and "time constants"—what he says may have a great deal to do with production, sales and profits. These expressions refer to the ratio between the volume of material held in a processing plant at any instant and the rate of production. The "hold-up" of material may have great significance, even in terms of minutes or hours. Too much hold-up means unnecessary capital outlay for the processing equipment. At the other extreme, a small amount of hold-up with a tremendous through-put may mean that too much is being spent on the control system. We cannot build the plant first and hang the controls on later and expect the best results. What we need is the best compromise between all plant and no control, and all control and no plant.

What about the men who will operate and maintain these plants? *Even in the most robotized of the automatic factories there will be many men, and they will have interesting and responsible jobs.* They will be freed from the tiring, nerve-racking or even boring jobs of today's mass manufacturing. To win this freedom, however, they will have to upgrade themselves in skill and sophistication. The new controllers and instruments will call for a higher level of precision of repair and maintenance. A $50,000 controller cannot be hit with a hammer if the shaft doesn't fit into the hole on the first try. Men who have heretofore thought of electronics equipment as merely a metal chassis with tubes will become conversant with switching, flip-flop, peaking and other circuits. They will have to judge when to repair and when to throw away rather than stop production. We have here a paradox: today we cannot afford *not* to have lots of control, because a half-day shutdown of a plant may mean a $100,000 loss in potential sales.

These robots are not hurting the workman—they merely coax him none too gently into taking more responsible jobs, making bigger decisions, studying and using his mind as well as his hands.

AN AUTOMATIC CHEMICAL PLANT

by Eugene Ayres

LAST YEAR ONE of the countries of Asia employed a U. S. contracting firm to design a modern oil refinery. The firm submitted a design for a $50 million plant, and it included the usual array of control instruments. After studying the plans, the officials of the country, which has an embarrassing surplus of manpower, asked the designers to eliminate all automatic controls from the plant. The country could provide any number of thousands of men to record measurements and to control processes, and it was prepared to pay the price of lower efficiency and poorer-quality products to create this opportunity for employment. The contracting firm gave sympathetic consideration to the request, but its engineers finally decided that under no circumstances could control instruments be eliminated from the design. It was not just a question of operating costs or efficiency: without suitable control instruments a modern refinery simply could not operate at all.

If the 50,000 control devices in the oil refineries of the U. S. should go "on strike," we would be faced with social disaster. The refineries would become lifeless industrial monuments. If we undertook to replace them with old-fashioned, manually operated refineries to supply our present motor-fuel needs, we would have to build four or five times as much plant, cracking and some other modern chemical processes would have to be eliminated, yields of motor fuel from crude petroleum would drop to a quarter of those at present, costs would skyrocket, and quality would plummet. Automobile engines would have to be radically redesigned to function with inferior fuel. And because of lower motor-fuel yields, we would need to produce crude petroleum several times as rapidly as we produce it now. Technology in refining would be set back to the early 1920s.

In a sense the prediction that the machine age is now about to be followed by the age of automatic control is an oversimplification. Mechanical controls came in with the very beginning of the machine age, and made it possible. What we are now seeing, and shall see increasingly in the future, is a rapid extension of such control. Machines, processes and combinations of machines and processes will become more and more automatic.

The role that automatic control already plays in technology can be seen by considering the unhappy situation of Great Britain. One measure of a country's industrialization is its capacity for consuming fuel. This capacity depends largely upon the quality of mechanization, which in turn depends upon automatic control. In fuel-consumption capacity Britain's industries have declined steadily, in relation to the rest of the world, ever since 1875. This has happened in spite of her abundant coal resources (until recently) and the fact that she has some of the world's best-informed technologists. Her industrial leaders in the main have failed to understand the philosophy of automatic control. Little more than a tenth of the textile looms in Lancashire, for example, are automatic. Lancashire textiles cannot compete in world markets, and a business depression is under way. Some other British industries show the same backwardness in application of automatic control. In the U. S., by contrast, the intense competition in cost, yield and quality has driven us steadily toward more and more automatization.

The petroleum industry is a leading example. Excluding taxes, the price of motor fuel today is little higher than in 1920, in spite of the shrinking dollar and rising costs, and almost the sole reason is improvement in process equipment made possible by automatic control. Automatization has reduced costs in two ways: by yielding more production per dollar's worth of equipment while the plant is running, and by reducing shutdown time. When a plant is not operating, fixed charges go on just the same. Ten years ago, when cracking units were reasonably well equipped with control devices, the units could be counted on to operate an average of 87 per cent of the time. Today they are more likely to operate 93 per cent of the time. About half of this recent six per cent gain has come from improvements in the control systems themselves—development of new devices and more intelligent use of old ones. Most of the other half has come from changes in operation and design made possible by improvement in automatic control.

When the gain in operating factor is translated into dollars, it seems that the extra cost of control systems and the changes in process equipment that the control systems have made possible, have in general been amortized in less than a year of plant operation—a performance that far surpasses normal investment in refinery equipment. And the gain in operating factor is only part of it: there has been a much more important gain in the production capacity of our refineries.

A refinery is a model case of the kind of manufacturing operation known as continuous process, in contrast to the discontinuous step-by-step operation of, say, an automobile assembly-line. Crude petroleum feeds in at one end of the plant, flows continuously through a series of treatment chambers and pipes, and pours out a variety of finished products at the other end. Such an operation is particularly amenable to automatic control. Let us look at a typical refinery as an example of the application of machine control to a continuous process.

It is a bewildering kind of factory, with metallic towers rising twenty stories high, hundreds of miles of pipe, and only an occasional modest building. A few lonely men wander about the spectral monster doing supervisory or maintenance tasks here and there. The plant is almost noiseless, all but devoid of visible moving parts. Despite its apparent inertness, however, the plant is throbbing with internal heat and motion. Every day a quarter of a million barrels of oil flow unobtrusively into its maw, and about as many flow out in the form of dozens of finished petroleum products—all profoundly and specifically altered by processing. Forty tons of catalyst are being circulated every minute of the day and night. Great volumes of chemicals are being consumed in processing, and greater volumes of chemical intermediates are being manufactured. Scores of unit processes are interlocked, with a meticulous balance of energy distribution.

The nerve center of this mechanical organism is the control room with its control panel. Here are ensconced the human operators—attendants upon the little mechanical operators of the plant. The human operators watch, they sometimes help or correct the instruments, but only occasionally do they take over the major part of operating responsibility. Barring emergencies, they take over completely only when the plant is starting up or shutting down—normally only about once a year in a catalytic cracking plant.

Some of the automatic control instruments are mounted on the control panel; some are tucked away behind it; most are located in the refinery yard near the jobs they have to do. On the control panel are many things—indicators of measurements, indicators of valve positions, indicators of settings of controllers, knobs for changing these settings, facilities for shifting from automatic to manual control, knobs for effecting manual control, alarms and safety devices, recordings of measurements for operation analysis, and recordings of measurements for accounting. The restful appearance of the control panel and the refinery is deceptive. Five hundred controllers, 400 motor-operated valves, 1,500 indicators and 800 recorders are in slight but significant motion at all times— like the steering wheel of a speeding motorcar on a straight road. But while the little compensatory movements (this way and that) of the steering wheel have only direction to control, instruments on a refinery panel board must control a hundred variables— many of them dependent upon others.

AN AUTOMATIC CHEMICAL PLANT

Indicating instruments are merely to be observed. Like the clinical thermometer, they call for action only when the indication is abnormal. Some little thing may go wrong with one of the controllers. Some unpredictable variation may occur in the flow of fluids or in the strength of materials. Corrosion may start a leak in a pipeline. A fuse may blow out in an electric circuit. A storm may cause structural damage. Like a physician, the human operator becomes sensitive to abnormalities and promptly seeks to apply corrective measures. All factors for correction are on the control panel.

Periodically the operators receive reports from analytical laboratories, where chemists are testing the composition and quality of the products. The operators must know how to make manual adjustment of controller "settings" to conform with analytical trends.

While the operator stands guard over his mechanical helpers, automatic controllers reciprocate by standing guard sleeplessly over the safety of the operator and the plant. Any out-of-the-way event touches off a visible or audible alarm, warning the operator of trouble to come. And automatic controls immediately set in motion the first necessary steps for safety, more quickly than the operator could.

Refineries and other mechanically controlled continuous processes have become so complex that control panels of the types used just a few years ago would today be too large and too difficult to learn to understand. Long steps have been taken in the direction of developing smaller instruments and bringing the pattern of the flow of materials and energy within the ready comprehension of the operator. One important step has been the design of graphic instrument panels in which instruments of greatly reduced size are placed at appropriate locations on a simplified flow diagram of the plant. Even with small instruments, however, such a diagram sometimes would be 100 to 200 feet long, so that compromises must be made to bring the instruments into more compact arrangement. Flow lines on the panel are given distinguishing colors to make interpretation as easy as possible. The diagram of the catalytic cracking process alone requires more than a dozen colors, for we are concerned with flows of several different systems and materials—hydrocarbons, water, steam, air, electric power, catalysts and other chemicals.

Now let us look into this maze and see what we are trying to control. If we could set up an unvarying set of conditions and depend on the system to operate constantly at the same rate of flow, temperature, pressure, and so on, we would need no control devices. But physical factors are imperfect. Mistakes are made by pumps as well as by people. The composition of the charge flowing into a process is likely to vary. Deposits may form in a valve. The heat losses from equipment will not be the same on cool nights as on hot afternoons. A shift in wind direction or velocity may alter the draft in a chimney.

We have two kinds of variables to control. First there are the conditions that determine the yield and quality of the product; these must be kept constant at optimum levels. Then we have another group of conditions which can be made to vary widely. The formidable job of the control devices is to keep changing the latter conditions in such ways as may be required to keep the determining conditions nearly constant. In other words, inanimate devices are charged with the responsibility of correcting inanimate mistakes.

The relationships among the variables are bewilderingly complex. Consider a fluid catalytic cracking plant. Essentially the process involves converting petroleum to gasoline and other products by heating it in the presence of a solid catalyst in the form of a fine powder; the catalyst makes it possible to operate at lower temperature and pressure than in simple thermal cracking. The main reactions take place in a large chamber called a "reactor," which is charged with a mixture of crude oil, steam and hot powdered catalyst. The heating in the reactor drives off oil vapors and steam to a fractionating column. The powdered cata-

AN AUTOMATIC CHEMICAL PLANT

lyst flows to a "stripper" for stripping off oil vapor trapped in it, then to a "regenerator" for burning off residual carbonaceous material, and finally is returned to the reactor to be used again. In the fractionating column gasoline fractions from the top of the column are subjected to various other fractionating operations to provide a product of desired boiling range, and the heavier oil in the column may be recycled for further breakdown.

Thus the catalytic cracking operation is a combination of several distinct processes. In a modern plant they are indissolubly linked together through control mechanisms. In the past 12 years about 70 fluid catalytic cracking units have been built in the U. S. Significantly, each unit has had more instrumentation than the one before. The earliest units had about 100; one of the most recent has about 500. Many of these are simply indicating or measuring instruments, but about 150 are for automatic control, and many of the control instruments also make records of measurements. A diagram showing all the linkages between the controllers and the process equipment looks as complicated as a blackberry bramble, and it takes a full-sized book to list the controllers with concise descriptions of their functions and linkages.

I can mention here only a few of the more interesting aspects of the control system in such a plant. One of the chief problems, of course, is control of the temperature. Part of the heat may be generated by combustion of fuel in a furnace. The rate of heat generation is one of the secondary conditions that may be made to vary within certain limits to compensate for changes in the flow of oil or for variations in heat losses. Now combustion itself has its own determining variables for efficiency. Naturally we want to get the maximum yield of heat from a given amount of fuel. Automatic control has increased combustion efficiency in a cracking plant roughly fivefold since 1930.

The temperatures of the components in the cracking process are all related to one another, and their nerve center is the reactor. The reactor temperature is one of the determining conditions for

47

the yield and quality of gasoline. It depends not only on the temperature of the preheated oil fed to the reactor but also, to a greater extent, on the temperature and the amount of hot catalyst recycled into the chamber. The reactor temperature is thus sensitive to the rate at which regenerated catalyst is recycled. This rate must be controlled by the reactor temperature. The amount of regenerated catalyst is another determining variable, but this is not controlled; at least one degree of freedom must be provided in any continuous system. The temperature of the regenerated catalyst is controlled largely by the rate of circulation of some of it through a cooler. The flow of air to the regenerator, for burning carbon off the catalyst, is varied to take care of combustion requirements—sometimes by the composition of the spent combustion gas.

Many other controllers are required for this part of the process. Some are in the steam system. Steam, generated by the catalyst cooler and by whatever other waste heat is available, forms part of the mixture flowing to the reactor and also serves to remove hydrocarbon material from spent catalyst before regeneration. All these operations require mechanized control to maintain thermal and material balance. The extensive air system requires controls to blow air at proper rates to proper points. The flow of recycled oil from the fractionating columns must be controlled. Spent catalyst must be replaced by fresh catalyst as fast as it deteriorates. These and many other details (such as intermediate storage-tank levels and gravity separators) require appropriate controls.

Compared with other continuous processes, fluid catalytic cracking is unique in several respects. Vapor containing abundant, finely divided solid catalyst acts like a liquid instead of a gas. For example, levels may be determined by pressure differentials. But to maintain the hydraulic properties of the mixture, it must be kept continuously in motion at a certain minimum rate. The automatic controls must operate massive equipment: the slide valves that control the flow of catalyst, for instance, sometimes weigh ten tons.

The fractionating towers present another set of control problems. The temperature and pressure at the top of the tower control the rate at which vapor is pumped into the tower. The level of liquid in the bottom of the tower (or in a "reboiler") is controlled by the outflow of oil recycled to the catalyst reactor and of oil sent to a "stripper" for separation of its light hydrocarbons. The temperature at the bottom of the tower is controlled by the temperature pressure conditions at the top or by the flow of charge or by both. Various side streams are withdrawn from the tower at controlled (usually constant) rates. Liquid levels in the "stripper" are controlled by rates of flow to storage and to other parts of the plant.

An interesting thing about fractionating-tower control is that its control instruments must be designed to function slowly and deliberately. They must not change secondary conditions more rapidly than the tower achieves equilibrium—otherwise the operation of the tower will be uneven or impossible. The tower attains equilibrium slowly because of its relatively large volume of liquid. Hence the tower-controls must respond to signals slowly and adopt for their slogan "easy does it."

Since 1920 oil refineries have been revolutionized by three basic advances in design, which have brought a sensational increase in production capacity per ton of equipment steel. All three were made possible only by automatic controls. The first was the replacement of the shell-still, in which large batches of oil were heated like water in a teakettle, by tubular heaters, in which oil is passed continuously and rapidly through pipes in a furnace. The tubular heating system, which speeded up the flow of oil and decreased the volume of liquid "in process," brought drastic reductions in operating hazards and costs. The second development was a change in furnace design to supply heat by radiation from flames or hot brickwork instead of depending upon actual contact with hot vapors of combustion. This brought about a sixfold increase in the rate of heat transfer—from 2,000 to 12,000 British thermal units per square foot per hour. The third advance was the improvement of furnace combustion efficiency that I have mentioned.

Within a decade, between 1930 and 1940, these contributions of automatic control doubled the plant capacity that could be built at a given cost. And aside from cost, improvements in automatic controls and consequently in equipment design have probably been responsible for the conservation of 100 million barrels of petroleum per year in the U. S.

Yet we have gone only part way along the road. There will be further improvements in the control devices and in the design of equipment to take advantage of them. And we are certain to see the one capping major advance that will make the refinery almost fully automatic—end-point control. What this means is simply a master controller which will continually analyze the end products, compute what changes must be made in the process conditions, and signal instructions back to appropriate points.

At first thought this may seem an unnecessary refinement. If the determining variables within the process are held constant by controllers at each point, will not the composition of the product be constant? In many cases it may. But for some processes the most important criterion is the concentration of some specific component in the product. An example is the catalytic cracking of distillate oil to get a maximum yield of butenes, needed for aviation fuel. To maintain the concentration of such desired products at constant optimum levels, we must have electronic instruments which can continually change certain secondary variables.

The petroleum industry is at the very beginning of development of such end-point control. Certain preliminary steps have been taken. Among the most important of these has been the development of methods of continuous stream analysis by infrared spectroscopy, mass spectrometry and X-ray photometry. Infrared methods are being applied to continuous determination of various olefines, carbon dioxide, carbon monoxide, sulfur dioxide and methane. The mass spectrometer can make almost continuous determinations of eight hydrocarbon components at once. The X-ray photometer can make continuous determinations of tetraethyl lead in the range of concentrations used in motor fuel.

The next step, now being prosecuted vigorously, is the incorporation of continuous analytical equipment in process control systems. Up to now there has been little incentive for any general application of end-point control to motor-fuel manufacture, because no one has succeeded in finding analytical criteria for the quality of motor fuel. But some petroleum processes are even now being controlled by a single property of the product, such as its viscosity, its refractive index, its density or its pH. And end-point control is being applied to supplementary motor-fuel processes such as catalyst regeneration, isobutane recovery and de-ethanization.

Automatic is a relative term. The so-called "push-button" factory, into which category a modern refinery falls, is of course not completely automatic—people still have to push the buttons. And even when button-pushing or valve-setting is replaced completely by impulses from end-point controllers involving continuous analysis, a refinery will still not be fully automatic. Some unpredictable variable is always likely to remain to tax the minds of human operators.

With all the help that science and art can give, refinery operators must possess a high order of intelligence. Automatic controls have not and never will substitute for intelligence. Indeed, they have raised the quality of personnel requirements to new high levels. The fact that robots can be devised to do nearly any routine operation performed by man, and much more besides, does not mean that our minds are to become less necessary. On the contrary, in a collective sense our minds are being taxed more seriously. The maintenance and operation of automatic systems calls for clarity of mind and technical intuition. And controllers have no intelligence except that which mathematical and engineering genius builds into them. The design of controlled systems to accomplish tasks that have no precedent in experience requires the utmost in imagination and inventiveness.

One of the most serious problems in automatic control has been the lack of men with an over-all understanding of the control system and the plant and process to which it is applied. This problem of management has retarded the progress of mechanization. There are many qualified specialists in instrumentation, but their specialization hampers communication and the integration of a plant. So someone has invented the term "system engineering" to include not only the control system but the plant itself. For in a very real sense the whole plant is a huge instrument, and the control devices are merely component parts. The system engineer must be distinguished in electronics, pneumatics, cybernetics—and common sense. The system engineer, if he can be developed, will have a great future.

Men have ever regarded the machine as a mixed blessing. Machine-made goods are to many a synonym for inferior goods. Will automatization remove the last drops of human creativeness and variety from man's products? Such a fear is ill-founded. To be sure, textiles made by the machine loom have lacked the traditional charm of those woven by hand. But the shortcomings of the machine are not inevitable. With modern automatic controls, machine-made textiles can approach the beauty and quality and individuality of the "hand-made." Machines can now be designed not only to surpass the regularities of careful hand manipulation but also to duplicate faithfully the irregularities of the artist's inspiration. Similarly, the immense standardized output of motor fuel from controlled refineries is approaching equality with the best "hand-made" laboratory fuels that technologists have been able to devise. It is becoming more and more practicable to translate subtle improvements in functional quality from the research laboratory to commercial production.

AN AUTOMATIC MACHINE TOOL
by William Pease

THE METAL-CUTTING industry is one field in which automatic control has been late in arriving. The speed, judgment and especially the flexibility with which a skilled machinist controls his machine tool have not been easily duplicated by automatic machines. Only for mass-production operations such as the making of automobile parts has it been feasible to employ automatic machinery. New developments in feedback control and machine computation, however, are now opening the door to automatization of machine tools built to produce a variety of parts in relatively small quantities.

The problem will be clearer if we first review briefly the history of machine tools and their relationship to manufacturing processes. The story begins in the last quarter of the eighteenth century. Prior to that time the tools of the millwright, as the machinist of that day was called, consisted chiefly of the hammer, chisel and file. His measurements were made with a wooden rule and crude calipers. His materials were prepared either by hand-forging or by rudimentary foundry casting. Crude, hand-powered lathes were already in existence, but they were used only for wood-turning or occasionally for making clock parts.

The first machine tool in the modern sense of the word was a cylinder-boring device invented in 1774 by John Wilkinson. Wilkinson is by no means as well-remembered as James Watt, but it was his invention that enabled Watt to build a full-scale steam engine. For ten years Watt had been struggling vainly to turn out a cylinder true enough for the job. After one effort he reported in discouragement that in his cylinder of 18-inch diameter "at the worst place the long diameter exceeded the short by three-eighths of an inch." But in 1776 Watt's partner, Matthew Boulton, was able to write: "Mr. Wilkinson has bored us several cylinders almost without error; that of 50 inches diameter, which we have put up at Tipton, does not err the thickness of an old shilling in any part." The importance of Wilkinson's boring machine cannot be overestimated. It made the steam engine a commercial success, and it was the forerunner of all the large, accurate metal-working tools of modern industry.

Another productive Englishman of the same period was Joseph Bramah. His inventions included one of the most successful locks ever devised, the hydraulic press, various woodworking machines, the four-way valve, a beer pump and the water closet. To manufacture his inventions he and an associate, Henry Maudslay, created several metal-cutting machines. The most significant of these was a screw-cutting lathe with a slide rest and change gears remarkably like our modern lathes.

The next great step forward in machine technology was pioneered by Eli Whitney. Although he is remembered mainly as the inventor of the cotton gin, his greatest contribution was an innovation of much more general import: interchangeability of manufactured parts. In 1789 Whitney, having made little money from the cotton gin, set up in New Haven as a manufacturer of muskets for the U. S. government. He employed on this contract the interchangeable system of manufacture which was at that time still considered impractical by most experts. In fact, two years later it was necessary for Whitney to go to Washington to reassure the Secretary of War and officers of the Army that his idea was sound. Displaying ten muskets, all tooled as nearly alike as he could make them, he showed that the gun parts could be interchanged among all ten without affecting the guns' operability. He went on to prove in his New Haven shop that precision machinery operated by relatively unskilled labor could make parts accurately enough for interchangeability, so that expensive handwork was no longer required.

Whitney's idea was received with skepticism, but it eventually won out. Interchangeable manufacture is a fundamental principle of all quantity production as we know it today. Automobiles and washing machines, typewriters and egg beaters—every common machine we use is manufactured with interchangeable parts.

The two primary tools of interchangeable manufacture are the lathe and the milling machine. The modern lathe owes its form mainly to Maudslay, but about 1854 the addition of the tool-changing turret equipped the lathe and its cousin, the automatic screw-cutting machine, for interchangeable manufacture. The first milling machine suitable for interchangeable manufacture was built by Whitney. In 1862 the Providence inventor Joseph R. Brown developed the universal milling machine, the type in common use today. To these basic tools the nineteenth century added machines for drilling, punching, sawing and shaping metal.

In a sense the guides, tracks and other devices built into machinery to raise its precision level are a kind of automatic control. This type of automatism is no new concept in the metal-cutting industry. From the beginning machine tools were created to reduce the amount of human skill required in manufacture. These automatic aids to proficiency, always adhering to the double principle of accuracy for interchangeability and speed for economy have increased through the years.

Flexible machines, capable of manufacturing a wide assortment of parts, are an essential part of modern manufacturing technique. The reason why they have thus far been untouched by automatic control can be given in terms of the concepts of information flow and feedback control.

A rough measure of the cost of automatic control is the amount and nature of the information the automatic machine must handle. To perform a complicated operation such as manufacturing a metal part, we must build into the machine a great deal of information-handling capacity, for it has to carry out a whole complex of instructions. This initial equipment is expensive. If the machine is to manufacture only a single product, say an automobile crankshaft, in large quantities, the investment is spread over many items and the cost of each crankshaft is small. In such a case the automatic machine is worth its cost.

But suppose we want an automatic machine which will make not one particular product, or part, but a number of different kinds of products, and only a few of each—as the versatile machine tool must do. Now the machine must handle a different set of instructions for each product, instead of the single set of instructions for the crankshaft. In other words, it must be able to deal with more information. And the cost of the information-handling capacity needed for each product is spread over only a few items instead of many. This is the essential problem in automatizing machine tools.

Obviously one way to attack the problem is to economize in the information requirements of the machine for the various operations. What are these requirements? To begin with, the machine tool must orient itself continuously toward the material on which it is working; if it is to drill a hole in a piece of metal, it must drill the hole in the right place and to the right depth.

Early in their history machine tools began to acquire automatic feed mechanisms. Maudslay's screw-cutting lathe, which controlled precisely the distance that the cutting tool was advanced for each revolution of the work piece, was an expression of this principle. Another step toward automatization was Thomas Blanchard's invention in 1818 of the "copying" lathe for turning gunstocks—the first of a class of tools now known as "cam-following" machines. This type of tool is automatically oriented to machine irregular shapes. The information required to specify the irregular shape is stored in a cam built to represent that shape. An important weakness of these machines is that all the force required to position the cutting tool is furnished by the cam itself. It is costly to build a mechanism strong and accurate enough to transfer motion from the cam to the work piece, and the cam wears out.

Feedback control made its first significant entrance into the

AN AUTOMATIC MACHINE TOOL

machine-tool field in 1921, when John Shaw, working in the shop of Joseph Keller, invented the Keller duplicating system. In this system the information source is not a cam but a plaster-of-paris or wood model of the part to be machined. An electrical sensory device traces the model shape and transfers the information to the tool. By permitting the use of soft, easily fabricated models, the method reduces the cost of information storage. Modern versions of the duplicating system are embodied in the die-sinking machines of the Pratt & Whitney Division of the Niles-Bement-Pond Company. An hydraulic form of it was originated by Hans Ernst and Bernard Sassen at the Cincinnati Milling Machine Company in 1930. Since World War II a number of manufacturers have developed systems of this kind, employing a variety of electrical, mechanical and optical devices.

A further step in the reduction of the cost of information storage and transfer is promised through the use of digital information processes. A number of applications of digital processes to machine-tool control are currently being made. Let us look in detail at one of the most ambitious of these completed this year at the Massachusetts Institute of Technology.

The M.I.T. system combines digital and analogue processes under feedback control to govern a milling machine whose cutting tool moves in three planes relative to the work piece. In this case the "model" of the object to be fabricated is supplied to the machine in the form of a perforated paper tape similar to that used in teletype systems. For a typical operation, 10 feet of tape will keep the machine busy for an hour.

The components of the M.I.T. system are grouped into two major assemblies. The first of these, called the "machine," comprises the milling machine itself, the three servo-mechanisms employed to operate its moving parts, and the instruments required to measure the relative positions of these parts. The second assembly, called the "director," contains all the data-handling equipment needed to interpret the information on the tape and to pass it on as operating commands to the machine. The director contains three major elements—a data-input system, a data-interpreting system and a set of three decoding servo-mechanisms.

The purpose of the data-input system is to take the original instructions off the tape and feed them into the interpretive and command elements of the director. It consists of a reader, whose metal fingers scan the tape and report the presence or absence of holes by electrical signals, and a set of six relay registers (two for each of the basic machine motions) which store and transmit this information in numerical form. The registers are supplied in pairs, so that while one of them is in control of the machine, the other can receive information from the tape. At the end of each operating interval, command is transferred instantaneously from one register to the other.

The data-interpreting system picks up the numerical instructions stored in the registers and transmits them as pulse instructions to the decoding servo-mechanisms. These pulses are generated by an electronic oscillator, the "clock," which acts as the master time reference for the entire system. By means of a series of flip-flop switches, and in accordance with the instructions stored in the registers, these pulses are sent on to each of the three decoding servo-mechanisms.

Up to this point in the process, information has been handled in digital form. The three servo-mechanisms now convert the instructions to the analogue form required by the machine tool. Pulses from the data-interpreting system are translated into the rotation of a shaft—one degree of rotation for each pulse. The shafts are connected to synchro transmitters which are themselves connected to the drive servo-mechanisms of the machine. A feedback circuit, inserted at this point in the director, makes certain that the conversion from digital to analogue information has been accurately carried out. It works as follows:

The mechanical element of each decoding servo-mechanism consists of the shaft connected to the synchro transmitter, a unit

AN AUTOMATIC MACHINE TOOL

called a "coder" and a small two-phase induction servo-motor with appropriate gearing (see diagram below). The coder generates a feedback signal in the form of one electrical pulse for each degree of shaft rotation. The number of these feedback pulses

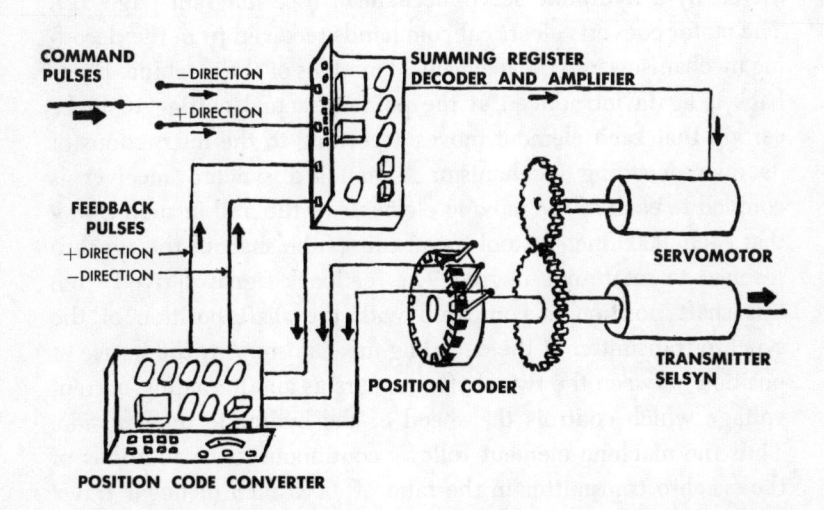

Decoding servo-mechanism ensures that commands are correctly translated from pulse form into the analogue form of the varying shaft angle. Command pulses from amplifier cause servomotor to rotate the transmitter shaft one degree per pulse. Rotation is sensed by brushes on position coder, which feed back one pulse to amplifier for each degree of rotation. The amplifier compares the number of feedback pulses with the number of command pulses and generates correcting pulses when errors are detected. Transmitter selsyn converts shaft motion into varying electrical signal which controls the machine-drive servo-mechanism diagrammed on page 61.

is then compared with the number of pulses emanating from the data-interpreting system by a device called a "summing register." If the two counts agree, the summing register is at zero; if they do not agree, an electric voltage is generated and the two-phase servo-motor rotates the shaft to bring the count to zero. Thus a feedback path makes certain that the output shaft position faithfully corresponds with the series of command pulses from the frequency divider. Information, coded first on tape, converted to digital and then to analogue form, is now transmitted to the working elements of the machine.

Each motion of the machine is accomplished by a lead screw driven by a hydraulic servo-mechanism (see diagram page 61). The motor converts electrical commands received from the decoding mechanisms into the mechanical motions of the machine. Feedback is again introduced at the point of actual cutting to make certain that each element moves according to the instructions of its own decoding mechanism. A standard synchro receiver is coupled to each of the moving elements of the mill in such a way that each .0005 inch of tool travel causes the shaft of the synchro receiver to rotate one degree. The feedback signal derived from this shaft position is compared with the shaft position of the synchro transmitter at the decoding mechanism. Any difference of position between the two shafts appears as an alternating-current voltage which controls the speed of the hydraulic transmission. Thus the machine element follows continuously the rotations of the synchro transmitter in the ratio of .0005 inch of linear travel to each degree of rotation.

How are the instructions for the machine's job put on the tape? The desired path of the cutting tool over the work is reduced to incremental straight-line segments; the segments are specified by numbers, and these are then translated into a code which can be punched on the tape.

The cutting path and the speed at which the work is to be fed to the machine are based on a number of factors: the amount of stock to be removed, the sequence of the machining operations, the setup of the work on the machine, spindle speeds, and so on. After the human operator determines the locus of the cutter center which will produce the desired cutting path, he divides the locus into a series of straight-line segments. They should be as

AN AUTOMATIC MACHINE TOOL

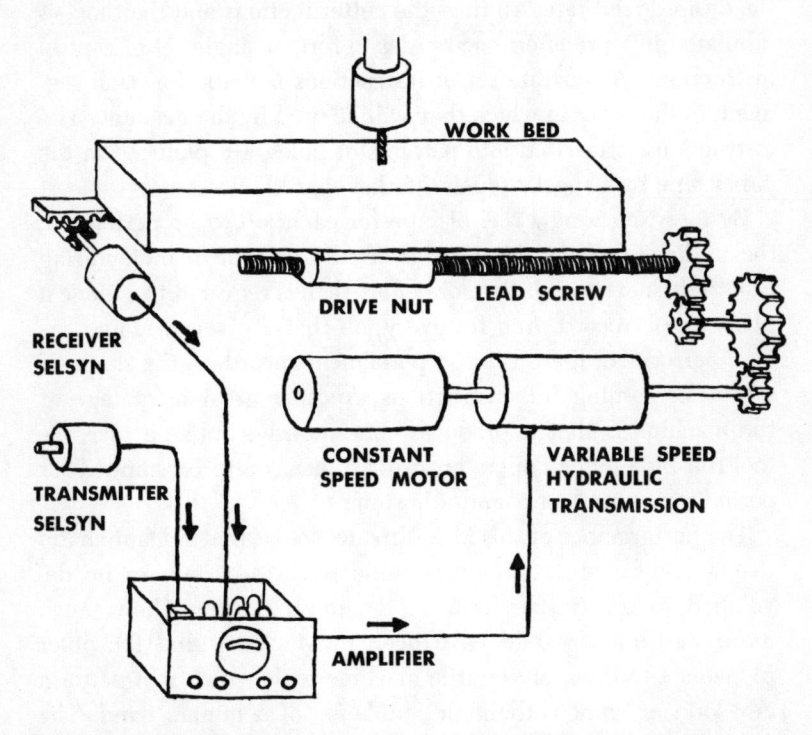

Machine-drive servo-mechanism ensures that commands in the form of a varying electrical signal from decoding servo-mechanism diagrammed on page 59 are correctly reproduced by the motions of the milling machine. The signal from the transmitter selsyn is relayed by the amplifier to the variable speed hydraulic transmission which drives the work bed of the milling machine. Receiver selsyn under the work bed at left feeds back to amplifier a signal corresponding to the motion of the bed. This signal is compared with the transmitter signal and a corrected signal is sent to the work-bed drive.

long as possible without differing from the ideal tool-center locus by more than the machining tolerance. The dimensions of each straight-line segment are then resolved into components parallel to each of the three directions of motion of the machine. For each straight-line segment, a time for execution is chosen to produce the desired feed rate. All this—the cutter motions and the time—is tabulated in a predetermined order to form a single set of control instructions. A separate set of instructions is made for each segment, in the order in which they will be used by the machine. The instructions, translated into patterns of holes, are punched in the paper tape by a special typewriter keyboard.

By inserting a new reel of tape for each job to be performed, the milling machine can be converted from one manufacturing task to the next with little more effort than is required to change a phonograph record. And for every job that a given machine has ever performed, there is left a permanent record, in the shape of a tape containing full instructions. Another great advantage of the machine is that it produces continuously; unlike a machine tool run by a human operator, it does not need to be stopped for periodic measurements and adjustments.

The performance of this M.I.T. model shows that full automatic machine tools are not only possible but are certain to be developed in practicable form. It is surely startling (how much more startling it would have been to Maudslay and the other pioneers!) to think of versatile machine tools which will perform any kind of work without the guidance of a human hand. The possible economic effects of such machines, on many industries besides metal-cutting, are beyond prediction. Automatized general-purpose machine tools, combined with high-production special-purpose tools, would make possible the automatic metals-fabricating factory. Nor are we restricted to metals. With digital machines in control we can conceive of factories which will process, assemble and finish any article of manufacture.

It is unlikely that the automatic factory will appear suddenly. Like the machine tool itself, it will just grow by steps until eventually it is here.

THE AUTOMATIC OFFICE
by Lawrence P. Lessing

WHEN THE FIRST giant, all-electronic digital computer went into operation in 1945, fantastic predictions about its capabilities were heard. Such "giant brains," it was said, would speedily take over all the paper functions of business and the running of entirely automatic factories. Actually these impressive monsters have proved harder to tame and put to work than was first thought. But their domestication is under way, and it is possible to report how some of them have begun to function on everyday business tasks.

The big machines are a different genus from the smaller electronic computers or data-processing machines that evolved out of them in some profusion after 1945. The small-scale electronic computers now widely used in business are employed only for limited routine tasks or special purposes; for example, one electronic computer handles all seat reservations at New York for a major airline, and another collates flight schedules for the Civil Aeronautics Authority. The jobs to be given to the new giants are of a higher order of magnitude. They embrace inventory control, general accounting, factory management and even the human equation.

Essentially these machines operate in the same way as the smaller types: they add, subtract, multiply and divide, and they work with "bits" of information represented by pulsed electric signals based on the binary number system. They have the same basic organs as any electronic computer: an arithmetical unit for computation, a control unit to direct the sequence of operations, a memory unit to store numbers and instructions, and various input-output devices for putting data and instructions into the machine and getting answers out. But they have many more of them, and their greater complexity enables them to work on very large problems, or groups of problems, without human intervention.

The giant computers still inspire the same awe that the first of the genus, ENIAC, did when its thousands of vacuum tubes began winking eerily in the basement of the Moore School of Engineering at the University of Pennsylvania nine years ago. There is something a bit frightening about them. An operator demonstrates their incredible speed by feeding in a problem involving some 780 arithmetical steps: before he has lifted his finger from the starting button the machine is typing out the answer. Another machine reads and translates Russian documents at the rate of 100 words a minute. Another takes in masses of personal data on college freshmen and predicts with better than 95 per cent accuracy which students will flunk out of the college course. (The same technique could be applied to the hiring of employees.)

ENIAC was designed to calculate ballistic tables and work on other complex equations in science and engineering. Scientific problems are still the main occupation of these machines, of which there are now more than thirty in the U. S. The reason their adaptation to business problems has been slow is not merely their cost ($1 million for a machine installed) but the fact that the kind of operation required for a business task is very different. A scientific problem usually involves a great deal of computation on relatively small amounts of data within a few fixed rules. Most business problems, on the other hand, are characterized by masses of data, relatively little computation and multitudes of variables. Attempts to translate these problems into large computer programs are just beginning.

The business possibilities are well illustrated, on a limited scale, by an inventory control machine called Distributon, which was built for the Chicago mail-order house of John Plain & Co. by

THE AUTOMATIC OFFICE

Engineering Research Associates, now a unit of the Sperry Rand Corporation. While this machine is by no means a "giant brain," it is the first major application of its kind and a simple approach to the larger type of problem. It is so simple that one of its designers says depreciatingly: "This machine does practically nothing, but it does nothing exceedingly well." Its first successful workout came in the Christmas rush just ended.

The problem was this. John Plain sells some 8,000 gift and houseware items by catalogue through about 1,000 retail merchants, mostly rural. Its business is highly seasonal, ranging from less than 2,000 orders per day in the off season to more than 15,000 per day in the Christmas period. The company must follow inventories closely and ship items fast when they are needed. For checking the inventories it employed a battery of women clerks, who recorded each order with a check mark against the catalogue number on a tally card and then registered the totals from hundreds of these cards each week on a master tally. Since the work was tedious and seasonal, it did not attract a high grade of workers. During the rush season the reports fell a week or two behind, and there were many errors.

The Distributon machine, with its 10 operators, replaced 60 tally clerks. The machine is about the size of an office locker. It contains a magnetic-drum memory and a small arithmetical-control unit for simple addition and subtraction, to which are attached 10 input units like small desk adding machines. The orders are recorded on a revolving magnetic drum, whose sensitized surface is divided into 130 invisible tracks capable of holding 39,000 digits or "bits" of information. The quantity of each item in stock is imprinted as tiny magnetized spots at a place designating the catalogue number, and the spots are arrayed at intervals on the tracks. When an order is received for a dozen, say of some item, the operator taps out the quantity and the catalogue number on her keyboard. If she types a wrong catalogue number, the machine flashes a signal and tells her to try again. The machine then searches the drum surface traveling at 100 miles per hour, finds the catalogue number, plucks off the stock total or the day's sales total, transmits it to holding relays, subtracts 12 from one total or adds 12 to the other and returns the new totals to the proper place on the drum—all in two-fifths of a second.

The machine can handle 90,000 tallies a day. To get out the sales total on any one item, an operator simply taps out 0 plus the catalogue number; the total instantly appears in illuminated numbers on a panel above her keyboard. For sales embracing more than one item the machine has a separate output unit in which a punched tape bearing the required catalogue numbers causes it to type out these numbers and their stored sales totals on a paper strip. Each night the machine is set to run off automatically a complete report on all the 8,000 items in the catalogue. To compile a daily stock report by the old tally-card method would have required some 150 clerks. Moreover, the Distributon does its work far more accurately than tally clerks could.

With modifications in size and in the type of input-output the same machine can handle many other kinds of inventory problems, materials scheduling and so on. It belongs to a class of moderately fast magnetic-drum machines which are now offered by a number of manufacturers. A neat, medium-sized model capable of a large variety of office jobs has reached quantity production by the Computer Research Corporation of California, recently acquired by the National Cash Register Company. The latest is the International Business Machines Corporation's Magnetic Drum Calculator. Somewhat more versatile than others, it stores up to 20,000 digits and up to 2,000 separate operating instructions. It can hold entire rate tables for calculating insurance policies, freight invoices, utility bills and the like, or entire production programs for scheduling the flow of parts and materials. The price range of the magnetic-drum machines is $50,000 to $100,000 or, if they are rented, from $500 to $1,000 a month. On a small scale they introduce into the business-machine field the important principle of the internal stored program.

THE AUTOMATIC OFFICE

What makes the big computers so revolutionary is the magnitude of their stored-program operations. The memory of such a machine has a capacity of millions of digits and thousands of instructions. Where the smaller electronic machines combine a few steps previously handled by other office machines or personnel, the big machines can combine many steps or take on a series of different problems in any sequence. Their price runs from $800,000 to $1 million installed, or about $10,000 a month on a rental basis.

Thus far only two organizations have built commercial machines of this scope. They are Remington Rand and International Business Machines. Remington Rand's giant computers are the UNIVAC (Universal Automatic Computer) which handles mass data, and the ERA 1103, a high-speed computer for engineering or industrial operations. IBM's creation is called 701; it is a large binary-digit machine designed chiefly for scientific-engineering work. Remington Rand got into the field by acquiring the Eckert-Mauchly Computer Corporation, whose founders designed both ENIAC and UNIVAC. Recently Remington Rand also absorbed the company that designed the ERA machine—Engineering Research Associates, Incorporated, a small firm which since 1944 has built more than 20 computer systems, mainly for the military. Six UNIVACs have been bought by government services, including the Census Bureau and Atomic Energy Commission, and this year's production calls for six more UNIVACs and six ERA 1103s. IBM is just rounding off production on 18 of its 701s, the first big computer to reach anything like mass production. It is expected to announce soon an advanced commercial version of this machine.

The few large-scale business problems which so far have been exercised upon these machines have reached them piecemeal. Some have been slipped into computer schedules at off times between engineering jobs. Others have run through central computer services set up by both manufacturers on a time-charge basis to educate business in their uses and to develop applications. The

67

problems run through so far begin to show the pattern of development.

For instance, Douglas Aircraft Corp., which has an IBM 701 computer that it has been using for tasks in engineering and aerodynamics, recently tried out the machine on a problem of manpower forecasting. Douglas had been offered several government contracts, each involving a somewhat different disposition of its engineering force. Given its staff of aerodynamic, structural, stress and other types of engineers, the timing and sequencing of their operations, what was the best order or combination of contracts to take on for the most advantageous use of manpower? This complicated problem would normally be solved, if at all, only by a mass of figuring, estimates and numerical projections. The computer, given all the relative data on manpower and contract specifications and instructions for their manipulation, came up with a set of schedule curves in less than an hour. This is a type of so-called "operational research" problem which involves so much data processing that most companies do not even attempt to work it out in detail.

The Monsanto Chemical Company recently took to IBM's computer center a problem in cost distribution. The task was to figure out how overhead charges, such as utility services, steam and the like, should be apportioned to calculate the cost of producing a certain chemical product. In multi-product industries, such as chemicals and oil, cost distributions become exceedingly complicated, and in many simpler industries they are so obscured that managers never get to know the actual cost per product in time for the information to have any significance. The problem that Monsanto presented to the 701 computer involved a large set of simultaneous equations and about 400,000 arithmetic operations. The machine worked out a cost sheet for the product in a few hours. Monsanto has ordered an advanced version of the 701 to prepare such cost sheets on about 1,200 items and to compute quarterly reports and do other accounting jobs which it has been doing for two years on much smaller IBM electronic computers.

THE AUTOMATIC OFFICE

In the same field of exploration the National City Bank of New York has run off a cost distribution on its various types of accounts and services. With banking costs constantly rising and clerical jobs hard to fill, a number of banks are considering these machines. Banks must keep permanent records of all transactions, must show on demand all operations performed on a depositor's account and must gear their accounts to the signature and amounts written on a depositor's check or deposit slip. Computers can handle the first two of these jobs. Magnetic tape and a recently developed high-speed printer can provide the permanent record. Closed-circuit television may take care of the showing of customers' accounts: the branch office of a New York bank is already trying this. For the problem of getting a computer to read the signature and the amount on checks and deposits no solution has yet been found, but the versatility the computers have shown suggests that even this may be accomplished.

The first real pulling together of piecemeal problems into one large computer system has been undertaken by the General Electric Company with the installation of a UNIVAC in its new Appliance Park center at Louisville, Ky., to handle all its factory bookkeeping. The idea behind Appliance Park is to centralize all of GE's now-dispersed electrical-appliance manufacturing units in one place, to gain savings in overhead, freight and other costs. Five separate departments have been assembled at Louisville and another located near by. To make so huge a complex feasible, a new order of centralized control is necessary, and no small part of the thinking behind it was the possibility of installing a giant computer before large clerical staffs were built up. GE has leased a UNIVAC from Remington Rand for two years, with an option to purchase.

The first assignment for UNIVAC is to take over various standard business and accounting jobs. It has started with the preparation of payrolls, which before had been handled on smaller electronic computers. Work data on several thousand employees are fed into the machine on magnetic tape. With instructions on wage rates, overtime, deductions and so forth stored in its memory, the machine calculates each worker's pay and then automatically types out payroll accounts and paychecks. It does the job in four hours. From payroll preparation, UNIVAC is to move on to certain routine clerical tasks, such as compiling cost distribution records, handling materials scheduling and inventory control for projecting the flow of production through the five departments, compiling sales records and preparing bills and, eventually, taking over the organization's complete cost and general accounting. When it has mastered these functions, the programing section plans to attempt something entirely new: rapid-fire sales analysis, fast enough to catch regional shifts in sales of various appliances and to modify production schedules accordingly.

Two insurance companies, Metropolitan Life and Franklin Life, will also soon put UNIVACs to work on actuarial problems, premium billings, preparation of dividend checks and other tasks.

The adaptation of the big computers to business operations is no minor task. Extremely complicated "breadboard" electronic mechanisms have had to be developed into reliable, more or less standardized production models. The diverse memory and input-output devices—more important in commercial uses than in scientific applications—have had to be worked up to rates of speed commensurate with the computer's internal millisecond arithmetic. The circuitry has had to be broken down into easily manufactured subassemblies which can be put together rapidly and swiftly replaced. Some of the machines check their own operations internally, flashing warning lights when anything goes wrong. Others self-check their marginal tubes (at abnormal power levels), picking off 99.9 per cent of the tubes that may go bad. All these techniques have brought the big computers' reliability up to about 98 per cent, which means, allowing for inspection and maintenance, that it runs effectively 85 per cent of the time. This reliability is as good as, or better than, that of punchcard machines, teletypes and other long-established equipment.

Business firms themselves will find it a big task to tailor their office procedures to a computer program. Most office practices have grown like Topsy, are untidy and are not known in all their details to any one member of the organization. Thus the first job is to find out exactly what is being done and to set up a staff group to program the flow of work for the computer. In one instance this preliminary phase alone took eight man-years of study. The clerical savings at this stage are elusive. Indeed, one company found that in the first months of putting a procedure into the computer, when it had to retain parallel operations in its old clerical-machine line, it went $200,000 into the hole on clerical costs.

Nevertheless, the promise of large data-handling machines is so great that big sums for exploration are justifiable. The U. S. Steel Corporation, for instance, in one of the most ambitious of programs, is investigating all possible applications of the new machines to the making and selling of steel. There have sprung up independent consulting services such as those supplied by Arthur D. Little, Incorporated, and Arthur Anderson, an accounting systems analyst, to lead businessmen through the mysteries and claims of electronics. For the new machines promise not merely clerical savings but higher managerial efficiency. Because of their lag in compiling needed statistics, industries have been forced to fly blind for varying periods. As more high-speed computers permeate commerce, operations will react more swiftly to actual conditions. Inventories and production will lose their tendency to pile up disruptively, decisions will be made on fuller knowledge. This knowledge must eventually yield a new type of national economy.

The slowness of development of these programs is cushioning the big computers' social impact. Eventually a machine of this kind takes over the routine work of hundreds of clerical workers. The automatic factory, a more difficult problem, is some distance away. But in paper work the computers will soon make possible a new level of speed and flexibility in the whole economy.

THE ECONOMIC IMPACT
by Wassily Leontief

APPROXIMATELY 500 YEARS ago the study of nature ceased to be solely a servant of philosophy and became a patron of applied arts and a source of practical invention. The economic development of the Western world has since proceeded at an ever-increasing pace; waves of technological change, driven by the surge of scientific discoveries, have followed one another in accelerated succession. The developmental lag between pure science and engineering application has progressively shortened. It took nearly 100 years for the steam engine to establish itself as part and parcel of the industrial scene, but electric power took less than 50 years and the internal combustion engine only 30. The vacuum tube was in almost every American home within 15 years of its invention, and the numerous progeny of Dr. Baekeland's synthetic plastics matured before we learned to pronounce "polyisobutylene." At the turn of the twentieth century it was said that "applied science is pure science 20 years later"; today the interval is much shorter—often only five years and sometimes but one or two.

From the engineering standpoint the era of automatic control has begun. Some of the fully automatic "factories of the future" are already on paper; they can be described and studied. Engineering, however, is only the first step; what automatic technology will mean to our economic system and our society is still decidedly a thing of the future. In judging its probable impact all we have to go by is tenuous analogy with past experience and theoretical deductions from our very limited information on the new techniques. And it is no help that some of the crucial facts and figures are veiled in secrecy.

THE ECONOMIC IMPACT

Important new inventions are traditionally held to presage the dawn of a new era; they also mark the twilight of an old. For some observers they contain promise; for others, fear. James Hargreaves constructed the first practical multiple spindle machine in 1767, and one year later a mob of spinners invaded his mill and destroyed the new equipment. The economists of the time (the golden age of "classical economics" was about to begin) came to the defense of the machines. They explained to labor that the loss of jobs in spinning would be compensated by new employment in machine-building. And for the next hundred years England did indeed prosper. Its labor force expanded both in textiles and in textile machinery, and wage rates by the end of the nineteenth century were at least three times as high as at its beginning.

But the men-v.-machines controversy blazed on. Karl Marx made of "technological unemployment" the cornerstone of his theory of capitalist exploitation. The conscientious John Stuart Mill came to the conclusion that, while the introduction of machinery might—in most cases would—benefit labor, it would not necessarily do so always. The answer depended on the circumstances of the case. And today that is still the only reasonable point of view one can maintain.

We are hardly in a position to reduce to detailed computation the effects that automatic technology will have on employment, production or our national standard of living. Aside from the paucity of our information on this new development, our understanding of the structural properties of our economic system itself is still incomplete. We must therefore rely on reasonable conjecture.

The economy of a modern industrial nation—not unlike the feedback mechanisms discussed throughout this volume—must be visualized as a complicated system of interrelated processes. Each industry, each type of activity, consumes the products and services of other sectors of the economy and at the same time supplies its own products and services to them. Just as the operating properties of a servo-mechanism are determined by the technical characteristics of the measuring, communicating and controlling units of which it is composed, so the operating properties of an economy depend upon the structural characteristics of its component parts and on the way in which they are coupled together. It is not by coincidence that in some advanced phases of his work the modern economist resorts to systems of differential equations similar to those used by the designers of self-regulating machinery.

The services of labor constitute one important set of inputs into the national economy. That it is the largest one is reflected in the fact that labor receives in wages some 73 per cent (in 1950) of the nation's annual net product. But labor is not the only type of input that goes into all other sectors. Certain natural resources, machinery, equipment and other kinds of productive capital feed into almost every branch of agriculture, manufacture, transportation and distribution. The distribution of our gross national product as between wages and salaries, on the one hand, and nonlabor income (profits, interest, rent and so on), on the other, tends to be stable from year to year. Since 1880, however, labor's share has steadily gained. Behind these figures lie the intricate processes of our economic development, influenced by such factors as population growth, the discovery of new and the exhaustion of old natural resources, the increase in the stock of productive plant and equipment and, last but not least, a steady technological progress.

A clear insight into the nature of that progress is to be found in the downward trend in the number of manhours required for an average unit of output. If we set the index at 100 in the year 1880, it has fallen today to less than 25. In the first 30 years of the period, the saving of labor seems to have been accompanied by a corresponding increase in capital investment. Between 1880 and 1912 the amount of machinery and of other so-called fixed investment per unit of output rose by 34 per cent, while the manhour input fell 40 per cent. Then the ratio of investment to output began to drop. We introduced more efficient machinery rather than just a greater quantity of it. That it actually was more efficient can be seen from the fact that labor productivity rose apace. In 1938 a unit of output consumed only about half as many manhours as would have been spent upon its production in 1918.

Such is the stage which the new technology—the technology of automatic control—has now entered. The best index we have of how far automatization has gone is the annual U. S. production of "measuring and controlling instruments." After hesitation during the depression and war years, instrument production is now rising rapidly. In part this rise mirrors the recent accelerating pace of industrial investment in general. But the real advance in instrumentation is indicated by the fact that the instrument production curve is rising far ahead of plant investment as a whole. A breakdown of the relative progress of automatic operation in individual industries shows that the chemical and machinery industries lead; next come metal processing (mainly in the smelting department) and ceramics. In interpreting these trends, however, one must bear in mind that instrumental control is less costly for some processes than for others.

The estimated cost of complete instrumentation of a new modern plant to automatize it as fully as possible today ranges from 1 to 19 per cent (depending on the industry) of the total investment in process equipment. The average for all industries would be about 6 per cent. On this basis, if all the new plants built in 1950 had been automatized, some $600 million would have been spent for measuring and control instruments. Actually the production of such instruments in 1950 totaled only $67 million. In other words, to automatize new plants alone, to say nothing of those already built, would require nearly 10 times as great an investment in instruments as we are now making.

Yet 6 per cent is far from a formidable figure. Furthermore, the investment in instruments would not necessarily mean a net increase in total plant investment per unit of output. On the contrary, the smoother and better-balanced operation of self-regulating plants has already shown that they can function with less capitalization than a nonautomatic plant of identical capacity. And much existing equipment can readily be converted from manual to automatic control. It therefore seems that the automatization of our industries, at least to the extent made possible by present technology, is likely to advance rapidly. The mechanization of the nineteenth century required heavy capital investment and proceeded slowly; the new technology, unhampered by such vast capital requirements, can be introduced at a much faster pace.

In transportation and agriculture, machines by now have practically eliminated the need for human muscle power. Man has all but ceased to be a lifter and mover and become primarily a starter and stopper, a setter and assembler and repairer. With the introduction of self-controlled machinery, his direct participation in the process of production will be narrowed even further. The starter and stopper will disappear first, the setter and assembler will go next. The trouble-shooter and repairman of course will keep their jobs for a long time to come; the need for them will even increase, for the delicate and complicated equipment of automatic control will require constant expert care. We shall continue to need inventors and designers, but perhaps not many even of them: the chief engineer of a large electronic equipment firm recently expressed to me his apparently well-founded hope that before long he would have circuits designed by an electronic machine, eliminating human errors.

All this inevitably will change the character of our labor force. The proportion of unskilled labor has already declined greatly in recent decades; it is down to less than 20 per cent. Meanwhile the numbers of the semiskilled have risen, and they now constitute over 22 per cent of the labor force. This trend has slowed down during the past decade, however. Now we shall probably see an accelerated rise in the proportion of skilled workers, clerks and professional personnel, who already make up 42 per cent of our working population.

In a country with a less fluid and more differentiated social structure than ours, these rapid changes in occupational composition of the population might have brought about considerable strain. But the celebrated, and often criticized, uniformity of American living renders the effects of such transition almost imperceptible. For example, recent studies indicate that the family of a typical $3,000-a-year clerk spends its money in very much the same way as the family, say, of a machine-press operator with a similar income.

Will the machine-press operator be able to earn his $3,000 when an automatic controlling device takes over his job? The answer must depend in part on the speed with which the labor force is able to train and to retrain itself. If such upgrading were to fall behind the demand of the changing technology, semiskilled and unskilled workers certainly would suffer unemployment or at least sharply reduced earning power. The experience of the last twenty years, however, has underlined the flexibility of U. S. workmen. Under the stimulus of the general American striving toward social and economic betterment, they have been quick to take to vocational training for new jobs. There has been no surplus of unskilled and semiskilled labor; indeed, wages in these fields have risen even faster than skilled and professional work.

But if automatic machines largely take over our production, will there be enough jobs, skilled or otherwise, to go around? Admittedly the possibility of eventual unemployment cannot be excluded on *a priori* grounds. If the capital investment were to increase rapidly while the need for manpower dropped, the resulting rise in capital's share of the national income could cause drastic unemployment. But as we have seen, the amount of capital needed for each unit of output has actually been reduced in recent years, and the installation of automatic machinery will further reduce it. Therefore labor should be able to maintain or improve its relative share of the national income. The danger of technological unemployment should be even smaller in the foreseeable future than it was at the end of the nineteenth century, when capital requirements were rising.

While the increase in productivity need not lead to involuntary idleness, it certainly does result in a steady reduction in the number of years and hours that an average American spends at making his living. The average work-week has been shortened from 67.2 hours in 1870 to 42.5 hours in 1950. This reflects a deliberate decision by the American people to enjoy an ever-increasing part of their rising standard of living in the form of leisure. If we had kept to the 67-hour week, we would be turning out a considerably greater amount of goods than we actually are. In other words, to enjoy shorter hours and longer vacations, we have deliberately chosen *not* to produce and hence not to consume all the commodities and services we could be producing by 1870 working-day standards. In fact, we have chosen to spend more and more of our ever-increasing production potential on leisure. The temporary shift to a high output of material goods during the last war only emphasizes this tendency, for we returned to the long-run trend immediately after the war. In the future, even more than in the past, the increased productivity of the American economy will be enjoyed as additional leisure.

Looking back, one can see that 1910 marked the real turning point in this country's economic and social development. That was the year when the last wave of immigration reached its crest; the year, also, when our rural population began to decline in absolute terms. Between 1890 and 1910 our national input of human labor had shot up from 28.3 million standard man-years to 42.5 million. Then in 1909 the model-T Ford began to roll off the first continuous production line. This great shift to mass production by machine was immediately reflected in shorter hours. In the next decade our manpower input increased by only one million man-years, and after 1920 it leveled off and remained almost constant until the early 1940s. Even at the peak of the recent war effort our total labor input, with an enormously larger population, was only 10 per cent greater than in 1910. Automatization will accelerate the operation of forces which have already shaped the development of this country for nearly half a century.

The new technology will probably have a much more revolutionary effect on the so-called underdeveloped countries than on the U. S. or other old industrial nations. Shortage of capital and lack of a properly conditioned and educated labor force have been the two major obstacles to rapid industrialization of such backward areas. Now automatic production, with its relatively low capital and labor requirements per unit of output, radically changes their prospects. Instead of trying to lift the whole economy by the slow, painful methods of the past, an industrially backward country may take the dramatic shortcut of building a few large, up-to-date automatic plants. Towering up in the primitive economy like copses of tall trees on a grassy plain, they would propagate a new economic order. The oil refineries of the Near East, the integrated steel plant built after the war in Brazil, the gigantic fertilizer plant recently put into operation in India—these are examples of the new trend in underdeveloped regions of the world. How formidable the application of modern technology in a backward country may become is demonstrated by the U.S.S.R.'s recent great strides in industrialization.

At the outbreak of the First World War the U. S. suddenly lost its source of many indispensable chemicals in Germany. Domestic production had to be organized practically overnight. The newly created U. S. chemical industry had no force of experienced chemical craftsmen such as Germany had. The problem was solved, however, by the introduction of mechanization and automatization to a degree theretofore unknown. The American plants were run with amazingly small staffs of skilled workers. The same thing is now happening, and possibly will continue on a much larger scale, in backward countries. Advanced design, imported mostly from the U. S., will compensate at least in part for their scarcity of high-quality labor.

Naturally automatization, while solving some problems, will everywhere create new and possibly more difficult ones. In Western civilization the liberation from the burdens of making a living has been going on for some time, and we have been able to adjust to the new situation gradually. In the rising new countries economic efficiency may at least temporarily run far ahead of progress toward social maturity and stability. Much of the stimulus for the educational advancement of the Western nations came from economic necessity. Automatization may weaken that powerful connection. It remains to be seen whether the backward countries will find a driving force to help them develop the social, cultural and political advances necessary to help them cope with the new economic emancipation.

PART 3 INFORMATION: THE LANGUAGE OF CONTROL

I. WHAT IS INFORMATION?
by Gilbert King

Gilbert King is a physical chemist who makes versatile use of the mathematical discipline of his specialty in such diverse fields as computer design, operations research and the quantum mechanics of chemical reactions. Born in England, he got his undergraduate and graduate degrees at Massachusetts Institute of Technology. King is now with International Telemeter Corporation, engaged in developing a metering system for subscription television.

II. THE MATHEMATICS OF INFORMATION *by Warren Weaver*

As director of the Division of Natural Sciences and Agriculture of the Rockefeller Foundation since 1932, Warren Weaver has played a pivotal role in the recent history of U. S. science, discovering and backing some of the most fruitful movements in research and the men who have carried them forward. He was born in Wisconsin in 1894, took his A.B. and Ph.D. at the state university and rose in its faculty to chairmanship of the mathematics department. During World War II, he organized and led the extraordinarily productive Applied Mathematics Panel of the National Defense Research Committee and himself was awarded the Medal of Merit for having "revolutionized" antiaircraft fire control. Weaver has taken the lead in the recent reorientation of the American Association for the Advancement of Science toward concern with the broad questions of the unity of science and its relations to society and this year is chairman of its Board of Directors.

III. INFORMATION MACHINES
by Louis N. Ridenour

During the war, Louis N. Ridenour was assistant director of the Radiation Laboratory at Massachusetts Institute of Technology, the center for the development of radar. As a physicist who had not been engaged in the development of the A-bomb and so had not had access to classified information surrounding it, he took a leading role in the first efforts to acquaint the public with the underlying scientific concepts and information. During the past decade, he has held posts as professor of physics at the University of Pennsylvania, dean of the sciences at the University of Illinois and chief scientist of the U. S. Air Force. He is now in industry, working on missile systems development at the Lockheed Aircraft Corporation.

WHAT IS INFORMATION?
by Gilbert King

The "lifeblood" of automatic control is information. To receive and act on information is the essential function of every control system, from the simplest to the most complex. It follows that to understand and apply automatic control successfully we must understand the nature of information itself. This is not as simple as it may seem. Information, and the communication of it, is a rather subtle affair, and we are only beginning to approach an exact understanding of its elusive attributes.

Think of a thermocouple that records the temperature of a furnace. The instrument translates the temperature into a voltage. This information seems straightforward enough. But as soon as we put it into a practical feedback loop to control the furnace temperature, we discover that the voltage signal is not a "pure" translation; it is contaminated by the heat due to random motion of the electrons in the thermocouple. The contamination is known as "noise." If we want to control the furnace temperature within a very small fraction of a degree, this noise may be sufficient to defeat our aim. In any case, the situation illustrates a fundamental property of information: in any physical system, it is never available without some noise or error.

Information can take a great variety of forms. In a thermocouple the voltage is a continuous signal, varying as the temperature varies. But information may also be conveyed discontinuously, as in the case of a thermostat, which either does or does not make an electrical contact. It gives one of two distinct signals —"on" or "off." The signals used in control may be numbers. The financial structure of the country is to a large extent controlled automatically (but not, as yet, mechanically) by the messages sent on ticker tape to hundreds of brokers, whose reactions affect the capital structure. Railway traffic is controlled by means of information transmitted on a teletype tape. Automatic control may include the human mind in the feedback loop, and in that case information takes the form of language messages, which may control the actions of people and nations. All literature, scientific or otherwise, represents messages from the past, and a literature search is a form of feedback loop which controls further thought and action.

During the past decade mathematicians have discovered with surprise and pleasure that information can be subjected to scientific treatment. Indeed, it meets one of the strictest requirements: it can be measured precisely. Information has been found to have as definite a meaning as a thermodynamic function, the nonpareil of all scientific quantities. In fact, as Warren Weaver shows in the next chapter, information has properties of entropy and may be treated as negative entropy. For the moment, however, we shall merely state that information is something contained in a message which may consist of discrete digits and letters or of a varying but continuous signal. Signals convey information only when they consist of a sequence of symbols or values that change in a way not predictable by the receiver.

Human beings have developed a number of systems, using sets of discrete symbols, for communication. These can be analyzed in quantitative terms. To keep track of a bank account of less than $1,000, for example, requires five "places" in a counting machine: units, tens and hundreds for dollars, units and tens for cents. Experience seems to have shown that 60 letters or places (12 words) are sufficient for most telegraph messages. In the decimal system we do all our counting with 10 digits; in the English language we use 26 letters. And there are many other sets of symbols, such as the dots and dashes of the Morse code, and so on.

WHAT IS INFORMATION?

In a control system, however, such sets of symbols may be too complex and cumbersome. The simplest type of components that we can use in a control loop is the kind of device (e.g., a relay or thyratron tube) that can assume only two states—"on" or "off." This means that it is most convenient to express message symbols on a binary scale, which has only two symbols: 0 and 1. Communication consists essentially in the progressive elimination and narrowing of the totality of all possible messages down to the one message it is desired to convey. If we visualize a recipient looking at a teletype awaiting the next symbol, we appreciate that each symbol reduces the number of possible messages by a factor proportional to the number of different symbols that might be sent. In the binary notation each symbol represents a simple choice between just two possible ones, and this has many advantages for expressing information.

In a message consisting of binary digits, each digit conveys a unit of information. From "binary digit" the mathematicians John Tukey and Claude Shannon have coined the portmanteau word "bit" as the name of such a unit of information. It is almost certain that "bit" will become common parlance in the field of information, as "horsepower" is in the motor field.

The number of bits in a message is a measure of the amount of information sent. This tells us exactly how much we are learning, and how much equipment is needed to handle the messages expected. Take as an example the recent suggestion that the contents of books be broadcast by television from a central library, thus doing away with the need for regional libraries. It takes seven bits to identify one letter or other character; on the average there are five letters in a word and 300 words on a page. Thus it would take only about 10,000 bits to transmit each page as a coded message. To televise a page, however, would require a great many more bits than that. In order to make the page legible, the screen would have to carry at least 250,000 black or white spots (corresponding to 500 lines vertically and horizontally). The image would have to be repeated 300 times, to allow the reader 10 seconds to read a page. Hence the required number of bits would be 75 million ($300 \times 250{,}000$) instead of 10,000. Since an increase in the amount of information sent requires an increase in the bandwidth of the broadcasting channel, it is clear that the televising of books is not an efficient method.

Can information in the form of a continuously changing voltage be of the same nature and be measured in the same units as numbers from a counting machine or words in a communication network? At first sight this does not seem possible, for it has long been considered axiomatic that a record of a continuous variable contains an infinite amount of significant information. Actually that is not so, for the reason that no physical measurement can resolve all of the information. The resolving power of a microscope, for example, is determined by its aperture, which is finite and therefore sets a limit on the fineness of discrimination. This theorem can be generalized to all instruments. Now Shannon, in his famous theorem on communication theory, has shown that when such a limitation exists, one can collect all the available information in a continuous signal by sampling it at certain finite intervals of time. Conversely, it can be proved that the continuous signal can be exactly reconstructed from the finite points, provided, of course, they are taken at the required frequency, determined by the aperture. A series of numbers, or of amplitudes sampled periodically, will completely specify the signal. Hence a message of this kind can be expressed as a series of binary digits.

So far we have used "message" and "information" interchangeably, but there is a distinction between them. The information content of the signals is reduced by the noise that comes with the message. The central problem of information theory, now undergoing investigation, is to determine the best methods of extracting the sender's message from the received signal, which includes noise. A magnetic storm can garble the telegram "I love you" into "I hate you." In fact, there is absolutely no way of being certain of transmitting a given message. Nothing is certain except chance.

One method of reducing the probability of error is to repeat the message. This does not improve the reliability very much and is expensive in bandwidth. The amount of "snow" in a television picture, for instance, could be halved by repeating each picture four times in the same interval of time, but this would demand four times the channel width. And the band available for television channels is limited and valuable. On the other hand, to get more information through a given bandwidth usually requires more hardware in the transmitter and the receiver, the amount increasing exponentially.

A more economical procedure for reducing the probability of error is to use redundancy. For instance, the message could be sent as "I love you, darling." This increases the chances of correct reception of the meaning without requiring as much extra time or bandwidth as mere repetition of the message would.

One of the cleanest examples of automatic control is the solution of a mathematical problem by such a procedure on a computing machine. A computing machine is a communications network in which messages (numbers) are sent from one part to another. The reduction of errors, naturally, is most important. However carefully the machine is constructed, errors inevitably creep in. Now usually there is no redundancy—the number 137 means one thing, and 138 distinctly another. But we can add redundancy, say by the method of carrying along the digit left after casting out nines, and can test these extra digits after each arithmetic operation. This requires more equipment (or bandwidth in a general sense), but a 20 per cent increase is sufficient to handle enough redundancy to reduce the possibility of overlooking an error to one part in 100 million.

The classic device for reducing extraneous noise in ordinary signals is a filter. For example, by cutting out high frequencies in a radio signal we can eliminate the high-pitched hissing components of noise without loss of message content, for the original message seldom contains such high frequencies. But to reduce the noise within the frequency range of the message itself is more difficult. Noise is universal and insidious, and elaborate devices are needed to overcome it. A wide variety of approaches, collectively called "filter theory," has been considered. The most fascinating is in the direction of suitable coding of messages with the aid of computers.

Let us examine a simplified illustration based on the problem of using radar for "Identification of Friend or Foe." We can send out a radar pulse of a specific pattern, which a converter in a friendly plane will change to another pattern but which an enemy plane will reflect unchanged. At long range, however, noise may confuse the pattern so that we cannot tell friend from foe. The question is: What is the most suitable pulse shape, and to what shape should it be converted, to give the smallest chance of making a mistake? Let us assume that the pulse wave can be above or below or at the zero level. We can therefore express the information in a ternary (instead of binary) notation of three digits: $-1, 0, 1$. We shall also assume (as often happens in practice) that a noise pulse of 1, added to a signal pulse of 1, gives 1 in our limited detection equipment. Now it is easy to see that if we merely used a positive pulse for "friend" and a negative pulse for "foe," one would frequently be converted into the other when the noise-to-signal ratio was high. Let us then introduce some redundancy by using a double pulse.

There are nine possible signals. They can be represented as vectors in two dimensions (see diagram on opposite page). Now of the nine vectors we need only two for our message. Which two should be chosen to give the least chance of error due to noise? It turns out that the best choice is the pair of vectors directly opposite to each other, because the noise patterns required to convert one to the other would in these cases be the least frequent.

Messages as a rule can be mapped with a great number of dimensions. And in such cases it proves to be feasible to select the

WHAT IS INFORMATION?

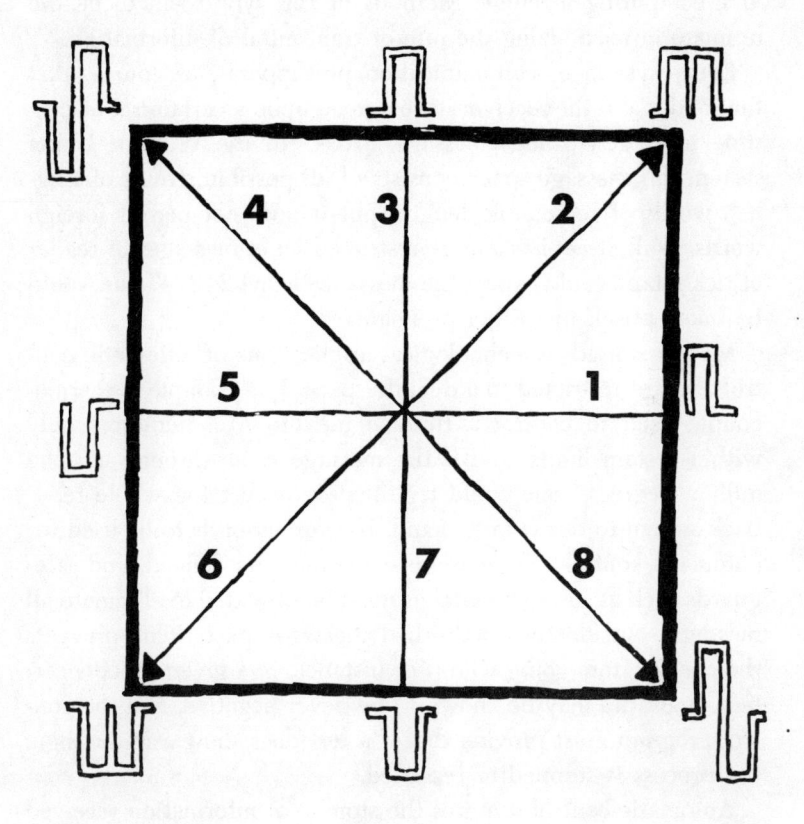

Simple code system shows how redundancy can help secure transmission of "yes" and "no." Pulse wave can be at, above or below zero— 1, 0 or -1. But noise might easily distort 1 to -1. With double pulse, as shown, 11 is less liable to be distorted to -1-1. Message 11 is like "yes, please"; -1-1 is like "no, thanks."

vectors entirely at random rather than by definite rules (which are too complicated to work out). Now a random signal, by definition, is noise. In other words, we have the paradox that the best way to encode a message is to send it as a typical noise pattern. The selected noise patterns correspond to an ambassador's code

89

book. The patterns can be decoded mechanically at high speed in a computing machine. Methods of this type seem to be the ultimate in maximizing the rate of transmittal of information.

Every system of communication presupposes, of course, that the sender and the receiver have agreed upon a certain set of possible messages, called "message space." In the Western Union system this message space consists of all possible strings of English words of reasonable length, but it does not permit foreign words. Wall Street has a more restricted set of messages. A reader of ticker tape could expect the message "Ethyl 24-1/4" but would be taken aback by "Ethel pregnant."

Messages used in technological applications of automatic control also are restricted to a definite space. For example, a thermocouple used to control a furnace measures temperatures only within certain limits, and if the message came through as "one million degrees," one could legitimately expect the whole feedback system to throw in its hand. If information is to be used for automatic control, the message space must be defined, and safeguards such as fuses or switches must be provided to eliminate all messages outside the established message space. This prevents the control from going wild. For instance, in a given process certain quantities may be known to be never negative, and the control program must provide that if a test does show a minus sign, the process is stopped or repeated.

Automatic control requires the storage of information received from the system's sensory instruments. For this a digital device, which simply stores numbers, is better than an analogue device. And the binary system is especially convenient. The most efficient known mechanism for the retention of information is the human brain. Recent physiological experiments suggest that the brain operates not with continuous signals but with sampled digital information, probably on a binary system; nerves seem to transmit information by the presence or absence of a pulse. The brain, with its ability to store vast amounts of information in a tiny space and to deliver specified items on demand, is the model which automatic control design strives to imitate.

Among artificial memory devices the most efficient is the photographic emulsion. Not only can it pack a great deal of information into a small area, but each spot is capable of recording about 10 distinguishable levels of intensity. Microfilm in particular is a very effective means of storing printed or pictorial information. Ultimately every man may have on microcards a library as large as he likes.

For the sake of compressing information into as small an area as possible, the ability of emulsions to record degrees of brightness is given up, and all that is asked of a grain is whether it is black or not. In other words, the technology of this medium is tending to a binary system.

The recording of information in the conventional form of printed matter is wasteful of space, even when the print is reduced by photography to microscopic dimensions. The printing of a letter requires a certain area of paper, which we can imagine as a grid with certain squares blackened to form the letter (see diagram on page 92). Some modern high-speed printers actually use this method, pushing forward certain pins from a matrix to mark each letter. In order to print passably legible letters the matrix must have at least 35 pins. In contrast, the binary digit notation needs only five places (instead of 35) to record the 26 letters of the alphabet, and only seven to give all the symbols of printing, including capitals, numerals and punctuation. Louis Braille recognized this when he used a binary system for his method of recording information for the blind.

The most efficient means of recording is by photography and binary digits. The finest commercial emulsion provides 32,400 resolvable dots or blanks per square millimeter. Allowing for the fact that at present the emulsion has to be mounted on a glass plate a millimeter thick, we have a medium which will store 40 million bits per cubic centimeter. If translated to binary code and

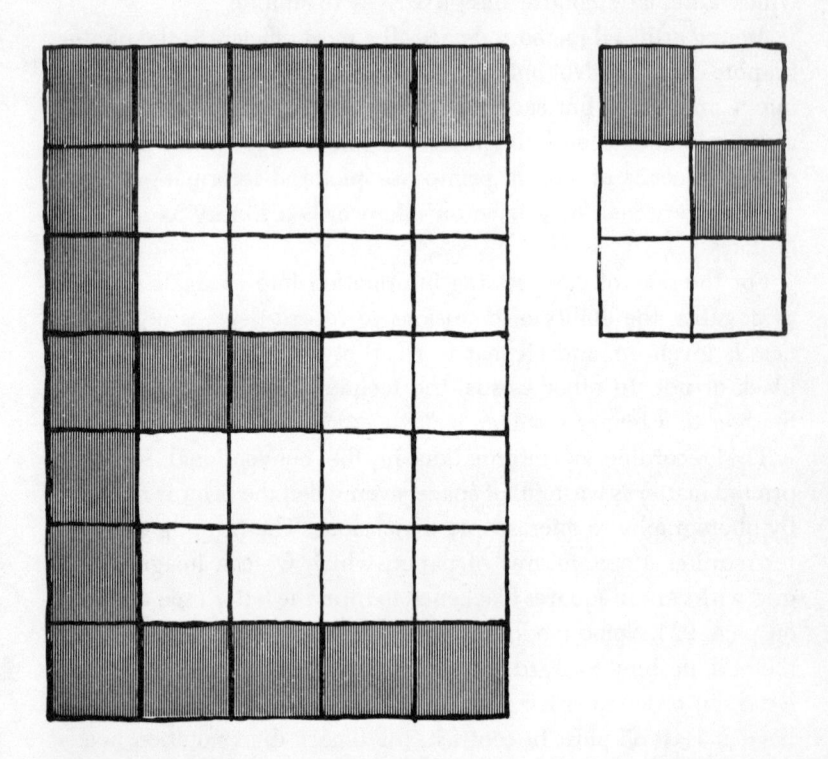

Letters need 35 bits (*here black and white squares*) for graphic reproduction. Braille uses but six bits.

recorded as black and white spots on this emulsion, all the words in all the books of the Library of Congress could be stored in a cubic yard.

Storage on photographic emulsion is not yet practicable because of the difficulties of retrieval: reading microfilm is not particularly easy or convenient. There are available, however, other means of compact storage—punched cards or tapes, magnetic tape or drums, electronic storage tubes, printed circuits with miniature tubes or transistors.

Automatic control requires storage of information for various purposes. In many cases the fact that information is stored is not apparent, but analysis shows that there is a delay during which the reported condition of the system is compared with certain standards. Discrepancies are discovered and corrected by feedback to the control organs. This is often done "instantaneously" by voltages stored in condensers, but more sophisticated control demands storage of a considerable history. We have seen that if a serious attempt is made to reduce noise associated with the message, sections of the received signal must be stored for a time to allow "filtering" of the signal by comparison with the code book.

This kind of control is "nonlinear," because the control signals are not simply proportional to the information supplied by the sensing instruments. Many aspects of these problems lie in new fields of applied mathematics, which when more thoroughly understood will open up wider fields of automatic control.

Some kinds of automatic control depend on statistical analyses of information received and stored in an extensive memory device. For example, electric-power companies have to control their production continuously as the load varies from hour to hour, day to day and season to season. A feedback loop from the immediate load is not sufficient, because stocks of fuel have to be accumulated in advance. The load must be anticipated many hours or even weeks ahead, and this can only be done with statistical knowledge. The power companies would be delighted to have good weather prediction, particularly for next winter, in order to increase their stockpile of fuel in ample time.

Another area where we can anticipate a substantial degree of automatic control based on statistical analyses is in the recently developed inventory machines used by stores. As each sale is made, a record is sent to a central computer, which in its most elementary form merely subtracts one from the inventory of the item sold. When the stock is reduced to a certain level, the machine prints instructions to replenish it.

This field of control comes under the head of operational analysis, which substitutes precise formulations for human opinion. It attempts to describe a business phenomenon by a working model, usually in the form of a set of equations. It is not difficult to visualize the possibility of using these formulas for automatic control of the routine part of a store's business. For example, the number of items to keep in stock depends on the probable sales, the profit and the cost of storage. After an operational analysis has yielded an answer to this complex problem, the solution can be inserted in an inventory type of computing machine as simple control numbers. A more complicated situation arises in judging when to fill stocks of several items which are interchangeable to a certain degree, e.g., shirts of different styles. The inventory machine now has to do a little arithmetical manipulation of the numbers of the various interchangeable items in its storage. A further elaboration comes when the inventory machine has to take into account the effect on sales of the onset of Christmas or of a competitor's sales promotion.

A working demonstration of the possibilities in this field is provided by the addressing machines now employed by large mail-order houses. Out of a list of, say, 15 million names, it has been customary to choose about half for the semiannual mailings of the catalogue. The selection of names, in the past, was made manually by rule-of-thumb procedures, such as any merchandiser might be expected to develop over a lifetime of experience. Today the names are processed on stencil plates in which a considerable variety of information bearing on the customer's prior sales history is encoded. With such information available to mechanical manipulation, it is possible to bring a high order of sophistication to bear upon the task of selection. In fact, fairly complicated equations are employed, with general economic and market trends factored into the weighting of the information on the stencils. The result has been to improve the net return on catalogue mailings by many millions of dollars.

We can expect to see in the future automatic machines which will make decisions in business and military operations by the application of the theory of games, developed by John von Neumann and others. One can readily imagine the inventory machines of two large department stores waging a battle for domination of a market. It is known, for example, that a sales campaign by one store sometimes stimulates sales in a competitor. One inventory machine may suspect, from a spurt in the sale of shirts by the store during an ordinarily quiet period, that the competitive store is in short supply of shirts. If it finds that its own store has a large stock of shirts, it will automatically suggest a sales campaign to put the competitor in an embarrassing position.

Explorations in the field of machine-devised game strategy are already being made. It is rumored that a computing machine in the U. S. will play a machine in England at chess. These machines will not attempt to play by arithmetic evaluations of all possible plays. They will have to learn to play the game and develop their own strategies. Learning is to a large extent the putting of information into a memory and the development of an ability to recognize connections. Appropriate or even ingenious actions may then be taken on the basis of the learned and stored experiences. Computing machines are capable of these processes.

The game-playing performances of machines will serve as valuable guides to the art of control in more important fields of endeavor. Research in science or engineering, for example, involves elements of automatic control. New experiments done in the laboratory, or new designs, are controlled by the success or failure of what has been done in the past. But the feedback is sadly incomplete. So much information is being produced that no research worker or engineer can be aware of all that is pertinent to his problem. In other words, the feedback loop is terribly congested. It is hoped that the new ideas of information theory will help to clear up the congestion and make the needed information more accessible.

We have seen that it is relatively easy to store library information for mechanical handling, say by recording on microfilm. The difficulty arises in locating quickly the information desired. Some progress has been made in this direction by Vannevar Bush and Ralph Shaw, who have developed a machine called the Rapid Selector. Printed matter is photographed on a large reel of microfilm in the order in which it is received. The information content of each frame is described by an abstract consisting of some 12 words. These words, called descriptors, are recorded in binary form as black and white spots on the margin of the frame. The coded words can be recognized by a photocell. When a research worker wants to extract from the microfilm all the material bearing on a certain item of information, he punches the descriptors defining this item on a card and inserts the card in the machine. The photocell searches the microfilm for these words at a speed of 5,000 pages a minute. Whenever the words on the card match the coded words alongside a frame, the machine makes a duplicate of the frame.

The Rapid Selector has not come into widespread use. One reason is that it takes time to assign descriptors to each page as it is microfilmed, and this constitutes a bottleneck. Another is that a mere dozen descriptors is too few to provide an adequate abstract. The recording problem might be made easier by using a reading machine to scan printed pages and translate their information into bits, which would then be recorded on the film. To retrieve information, however, a user would have to think of all the possible words and phrases that might have been used in describing the ideas for which he is searching. To design an efficient automatic library will require a good deal of thought and study.

Information is the most human of all the problems that the exact sciences have yet tackled. We shall need instruments like those of the human body—memory devices like the brain and control devices like the reflex networks in the nervous system—to handle information for automatic control. Progress is being made, and one of the most useful concepts developed so far is the "bit," by which information can be measured, stored, processed and transmitted most efficiently.

THE MATHEMATICS OF INFORMATION

by Warren Weaver

How do men communicate, one with another? The spoken word, either direct or by telephone or radio; the written or printed word, transmitted by hand, by post, by telegraph, or in any other way—these are obvious and common forms of communication. But there are many others. A nod or a wink, a drumbeat in the jungle, a gesture pictured on a television screen, the blinking of a signal light, a bit of music that reminds one of an event in the past, puffs of smoke in the desert air, the movements and posturing in a ballet—all of these are means men use to convey ideas.

The word communication, in fact, will be used here in a very broad sense to include all of the procedures by which one mind can affect another. Although the language used will often refer specifically to the communication of speech, practically everything said applies equally to music, to pictures, to a variety of other methods of conveying information.

In communication there seem to be problems at three levels: 1) technical, 2) semantic, and 3) influential.

The technical problems are concerned with the accuracy of transference of information from sender to receiver. They are inherent in all forms of communication, whether by sets of discrete symbols (written speech), or by a varying signal (telephonic or radio transmission of voice or music), or by a varying two-dimensional pattern (television).

The semantic problems are concerned with the interpretation of meaning by the receiver, as compared with the intended meaning of the sender. This is a very deep and involved situation, even when one deals only with the relatively simple problems of communicating through speech. For example, if Mr. X is suspected not to understand what Mr. Y says, then it is not possible, by having Mr. Y do nothing but talk further with Mr. X, completely to clarify this situation in any finite time. If Mr. Y says "Do you now understand me?" and Mr. X says "Certainly I do," this is not necessarily a certification that understanding has been achieved. It may just be that Mr. X did not understand the question. If this sounds silly, try it again as "Czy pan mnie rozumie?" with the answer "Hai wakkate imasu." In the restricted field of speech communication, the difficulty may be reduced to a tolerable size, but never completely eliminated, by "explanations." They are presumably never more than approximations to the ideas being explained, but are understandable when phrased in language that has previously been made reasonably clear by usage. For example, it does not take long to make the symbol for "yes" in any language understandable.

The problems of influence or effectiveness are concerned with the success with which the meaning conveyed to the receiver leads to the desired conduct on his part. It may seem at first glance undesirably narrow to imply that the purpose of all communication is to influence the conduct of the receiver. But with any reasonably broad definition of conduct, it is clear that communication either affects conduct or is without any discernible and provable effect at all.

One might be inclined to think that the technical problems involve only the engineering details of good design of a communication system, while the semantic and the effectiveness problems contain most if not all of the philosophical content of the general problem of communication. To see that this is not the case, we must now examine some important recent work in the mathematical theory of communication.

This is by no means a wholly new theory. As the mathematician John von Neumann has pointed out, the nineteenth-century Austrian physicist Ludwig Boltzmann suggested that some concepts of statistical mechanics were applicable to the concept of information. Other scientists, notably Norbert Wiener of the Massachusetts Institute of Technology, have made profound contributions. The work which will be here reported is that of Claude Shannon of the Bell Telephone Laboratories, which was preceded by that of H. Nyquist and R. V. L. Hartley in the same organization. This work applies in the first instance only to the technical problem, but the theory has broader significance. To begin with, meaning and effectiveness are inevitably restricted by the theoretical limits of accuracy in symbol transmission. Even more significant, a theoretical analysis of the technical problem reveals that it overlaps the semantic and the effectiveness problems more than one might suspect.

A communication system may be described summarily as consisting of a message source, a transmitter, a communication channel and a receiver. The information source selects a desired message out of a set of possible messages. (As will be shown, this is a particularly important function.) The transmitter changes this message into a signal which is sent over the communication channel to the receiver.

The receiver is a sort of inverse transmitter, changing the transmitted signal back into a message, and handing this message on to the destination. When I talk to you, my brain is the information source, yours the destination; my vocal system is the transmitter, and your ear with the eighth nerve is the receiver.

In the process of transmitting the signal, it is unfortunately characteristic that certain things not intended by the information source are added to the signal. These unwanted additions may be distortions of sound (in telephony, for example), or static (in radio), or distortions in the shape or shading of a picture (television), or errors in transmission (telegraphy or facsimile). All these changes in the signal may be called noise.

The questions to be studied in a communication system have

to do with the amount of information, the capacity of the communication channel, the coding process that may be used to change a message into a signal and the effects of noise.

First off, we have to be clear about the rather strange way in which, in this theory, the word "information" is used; for it has a special sense which, among other things, must not be confused at all with meaning. It is surprising but true that, from the present viewpoint, two messages, one heavily loaded with meaning and the other pure nonsense, can be equivalent as regards information.

In fact, in this new theory the word information relates not so much to what you *do* say, as to what you *could* say. That is, information is a measure of your freedom of choice when you select a message. If you are confronted with a very elementary situation where you have to choose one of two alternative messages, then it is arbitrarily said that the information associated with this situation is unity. The concept of information applies not to the individual messages, as the concept of meaning would, but rather to the situation as a whole, the unit information indicating that in this situation one has an amount of freedom of choice, in selecting a message, which it is convenient to regard as a standard or unit amount. The two messages between which one must choose in such a selection can be anything one likes. One might be the King James version of the Bible, and the other might be "Yes."

The remarks thus far relate to artificially simple situations where the information source is free to choose only among several definite messages—like a man picking out one of a set of standard birthday-greeting telegrams. A more natural and more important situation is that in which the information source makes a sequence of choices from some set of elementary symbols, the selected sequence then forming the message. Thus a man may pick out one word after another, these individually selected words then adding up to the message.

Obviously probability plays a major role in the generation of the message, and the choices of the successive symbols depend upon the preceding choices. Thus, if we are concerned with English speech, and if the last symbol chosen is "the," then the probability that the next word will be an article, or a verb form other than a verbal, is very small. After the three words "in the event," the probability for "that" as the next word is fairly high, and for "elephant" as the next word is very low. Similarly, the probability is low for such a sequence of words as "Constantinople fishing nasty pink." Incidentally, it is low, but not zero, for it is perfectly possible to think of a passage in which one sentence closes with "Constantinople fishing," and the next begins with "Nasty pink." (We might observe in passing that the sequence under discussion *has* occurred in a single good English sentence, namely the one second preceding.)

As a matter of fact, Shannon has shown that when letters or words chosen at random are set down in sequences dictated by probability considerations alone, they tend to arrange themselves in meaningful words and phrases (see table on next page).

Now let us return to the idea of information. The quantity which uniquely meets the natural requirements that one sets up for a measure of information turns out to be exactly that which is known in thermodynamics as entropy, or the degree of randomness, or of "shuffledness" if you will, in a situation. It is expressed in terms of the various probabilities involved.

To those who have studied the physical sciences, it is most significant that an entropy-like expression appears in communication theory as a measure of information. The concept of entropy, introduced by the German physicist Rudolf Clausius nearly 100 years ago, closely associated with the name of Boltzmann, and given deep meaning by Willard Gibbs of Yale in his classic work on statistical mechanics, has become so basic and pervasive a concept that Sir Arthur Eddington remarked: "The law that entropy always increases—the second law of thermodynamics—holds, I think, the supreme position among the laws of Nature."

1. Zero-order approximation
XFOML RXKHRJFFJUJ ZLPWCFWKCYJ FFJEYVKCQSGXYD QPAAMKBZAACIBZLHJQD

2. First-order approximation
OCRO HLI RGWR NMIELWIS EU LL NBNESEBYA TH EEI ALHENHTTPA OOBTTVA NAH BRL

3. Second-order approximation
ON IE ANTSOUTINYS ARE T INCTORE ST BE S DEAMY ACHIN D ILONASIVE TUCOOWE AT TEASONARE FUSO TIZIN ANDY TOBE SEACE CTISBE

4. Third-order approximation
IN NO IST LAT WHEY CRATICT FROURE BIRS GROCID PONDENOME OF DEMONSTURES OF THE REPTAGIN IS REGOACTIONA OF CRE

5. First-Order Word Approximation
REPRESENTING AND SPEEDILY IS AN GOOD APT OR COME CAN DIFFERENT NATURAL HERE HE THE A IN CAME THE TO OF TO EXPERT GRAY COME TO FURNISHES THE LINE MESSAGE HAD BE THESE.

6. Second-Order Word Approximation
THE HEAD AND IN FRONTAL ATTACK ON AN ENGLISH WRITER THAT THE CHARACTER OF THIS POINT IS THEREFORE ANOTHER METHOD FOR THE LETTERS THAT THE TIME OF WHO EVER TOLD THE PROBLEM FOR AN UNEXPECTED

Thus when one meets the concept of entropy in communication theory, he has a right to be rather excited. That information should be measured by entropy is, after all, natural when we remember that information is associated with the amount of freedom of choice we have in constructing messages. Thus one can say of a communication source, just as he would also say of a thermodynamic ensemble: "This situation is highly organized; it is not characterized by a large degree of randomness or of choice —that is to say, the information, or the entropy, is low."

We must keep in mind that in the mathematical theory of communication we are concerned not with the meaning of individual messages but with the whole statistical nature of the information source. Thus one is not surprised that the capacity of a channel of communication is to be described in terms of the amount of information it can transmit, or better, in terms of its ability to transmit what is produced out of a source of a given information.

The transmitter may take a written message and use some code to encipher this message into, say, a sequence of numbers, these numbers then being sent over the channel as the signal. Thus one says, in general, that the function of the transmitter is to encode, and that of the receiver to decode, the message. The theory provides for very sophisticated transmitters and receivers—such, for example, as possess "memories," so that the way they encode a certain symbol of the message depends not only upon this one symbol but also upon previous symbols of the message and the way they have been encoded.

We are now in a position to state the fundamental theorem for a noiseless channel transmitting discrete symbols. This theorem relates to a communication channel which has a capacity of C units per second, accepting signals from an information source of H units per second. The theorem states that by devising proper coding procedures for the transmitter it is possible to transmit symbols over the channel at an average rate which is nearly C/H, but which, no matter how clever the coding, can never be made to exceed C/H.

Viewed superficially, say in rough analogy to the use of transformers to match impedances in electrical circuits, it seems very natural, although certainly pretty neat, to have this theorem which says that efficient coding is that which matches the statistical characteristics of information source and channel. But when it is examined in detail for any one of the vast array of situations to which this result applies, one realizes how deep and powerful this theory is.

How does noise affect information? Information, we must steadily remember, is a measure of one's freedom of choice in selecting a message. The greater this freedom of choice, the greater is the uncertainty that the message actually selected is some particular one. Thus greater freedom of choice, greater uncertainty and greater information all go hand in hand.

If noise is introduced, then the received message contains certain distortions, certain errors, certain extraneous material, that would certainly lead to increased uncertainty. But if the uncertainty is increased, the information is increased, and this sounds as though the noise were beneficial!

It is true that when there is noise, the received signal is selected out of a more varied set of signals than was intended by the sender. This situation beautifully illustrates the semantic trap into which one can fall if he does not remember that "information" is used here with a special meaning that measures freedom of choice and hence uncertainty as to what choice has been made. Uncertainty that arises by virtue of freedom of choice on the part of the sender is desirable uncertainty. Uncertainty that arises because of errors or because of the influence of noise is undesirable uncertainty. To get the useful information in the received signal we must subtract the spurious portion. This is accomplished, in the theory, by establishing a quantity known as the "equivocation," meaning the amount of ambiguity introduced by noise. One then refines or extends the previous definition of the capacity of a noiseless channel, and states that the capacity of a noisy channel is defined to be equal to the maximum rate at which useful information (i.e., total uncertainty minus noise uncertainty) can be transmitted over the channel.

Now, finally, we can state the great central theorem of this whole communication theory. Suppose a noisy channel of capacity C is accepting information from a source of entropy H, entropy corresponding to the number of possible messages from the source. If the channel capacity C is equal to or larger than H, then by devising appropriate coding systems the output of the source can be transmitted over the channel with as little error as one pleases. But if the channel capacity C is less than H, the entropy of the source, then it is impossible to devise codes which reduce the error frequency as low as one may please.

However clever one is with the coding process, it will always be true that after the signal is received there remains some undesirable uncertainty about what the message was; and this undesirable uncertainty—this noise or equivocation—will always be equal to or greater than H minus C. But there is always at least one code capable of reducing this undesirable uncertainty down to a value that exceeds H minus C by a small amount.

This powerful theorem gives a precise and almost startlingly simple description of the utmost dependability one can ever obtain from a communication channel which operates in the presence of noise. One must think a long time, and consider many applications, before he fully realizes how powerful and general this amazingly compact theorem really is. One single application can be indicated here, but in order to do so, we must go back for a moment to the idea of the information of a source.

Having calculated the entropy (or the information, or the freedom of choice) of a certain information source, one can compare it to the maximum value this entropy could have, subject only to the condition that the source continue to employ the same symbols. The ratio of the actual to the maximum entropy is called the relative entropy of the source. If the relative entropy of a certain source is, say, eight-tenths, this means roughly that this source is, in its choice of symbols to form a message, about 80 per cent as free as it could possibly be with these same symbols. One minus the relative entropy is called the "redundancy." That is to say, this fraction of the message is unnecessary in the sense that if it were missing the message would still be essentially complete, or at least could be completed.

It is most interesting to note that the redundancy of English is just about 50 per cent. In other words, about half of the letters or words we choose in writing or speaking are under our free choice, and about half are really controlled by the statistical structure of the language, although we are not ordinarily aware of it. Incidentally, this is just about the minimum of freedom (or relative entropy) in the choice of letters that one must have to be able to construct satisfactory crossword puzzles. In a language that had only 20 per cent of freedom, or 80 per cent redundancy, it would be impossible to construct crossword puzzles in sufficient complexity and number to make the game popular.

Now since English is about 50 per cent redundant, it would be possible to save about one-half the time of ordinary telegraphy by a proper encoding process, provided one transmitted over a noiseless channel. When there is noise on a channel, however, there is some real advantage in not using a coding process that eliminates all of the redundancy. For the remaining redundancy helps combat the noise. It is the high redundancy of English, for example, that makes it easy to correct errors in spelling that have arisen during transmission.

The communication systems dealt with so far involve the use of a discrete set of symbols—say letters—only moderately numerous. One might well expect that the theory would become almost indefinitely more complicated when it seeks to deal with continuous messages such as those of the speaking voice, with its continuous variation of pitch and energy. As is often the case, however, a very interesting mathematical theorem comes to the rescue. As a practical matter, one is always interested in a continuous signal which is built up of simple harmonic constituents, not of all frequencies but only of those that lie wholly within a band from zero to, say, W cycles per second. Thus very satisfactory communication can be achieved over a telephone channel that handles frequencies up to about 4,000, although the human voice does contain higher frequencies. With frequencies up to 10,000 or 12,000, high-fidelity radio transmission of symphonic music is possible.

THE MATHEMATICS OF INFORMATION

The theorem that helps us is one which states that a continuous signal, T seconds in duration and band-limited in frequency to the range from zero to W, can be completely specified by stating 2TW numbers. This is really a remarkable theorem. Ordinarily a continuous curve can be defined only approximately by a finite number of points. But if the curve is built up out of simple harmonic constituents of a limited number of frequencies, as a complex sound is built up out of a limited number of pure tones, then a finite number of quantities is all that is necessary to define the curve completely.

Thanks partly to this theorem, and partly to the essential nature of the situation, it turns out that the extended theory of continuous communication is somewhat more difficult and complicated mathematically, but not essentially different from the theory for discrete symbols. Many of the statements for the discrete case require no modification for the continuous case, and others require only minor change.

The mathematical theory of communication is so general that one does not need to say what kinds of symbols are being considered—whether written letters or words, or musical notes, or spoken words, or symphonic music, or pictures. The relationships it reveals apply to all these and to other forms of communication. The theory is so imaginatively motivated that it deals with the real inner core of the communication problem.

One evidence of its generality is that the theory contributes importantly to, and in fact is really the basic theory of, cryptography, which is of course a form of coding. In a similar way, the theory contributes to the problem of translation from one language to another, although the complete story here clearly requires consideration of meaning, as well as of information. Similarly, the ideas developed in this work connect so closely with the problem of the logical design of computing machines that it is no surprise that Shannon has written a paper on the design of a computer that would be capable of playing a skillful game of chess. And it is of further pertinence to the present contention that his paper closes with the remark that either one must say that such a computer thinks, or one must substantially modify the conventional implication of the verb "to think."

The theory goes further. Though ostensibly applicable only to problems at the technical level, it is helpful and suggestive at the levels of semantics and effectiveness as well. The formal diagram of a communication system can, in all likelihood, be extended to include the central issues of meaning and effectiveness.

Thus when one moves to those levels it may prove to be essential to take account of the statistical characteristics of the destination. One can imagine, as an addition to the diagram, another box labeled "Semantic Receiver" interposed between the engineering receiver (which changes signals to messages) and the destination. This semantic receiver subjects the message to a second decoding, the demand on this one being that it must match the statistical semantic characteristics of the message to the statistical semantic capacities of the totality of receivers, or of that subset of receivers which constitutes the audience one wishes to affect.

Similarly one can imagine another box in the diagram which, inserted between the information source and the transmitter, would be labeled "Semantic Noise" (not to be confused with "engineering noise"). This would represent distortions of meaning introduced by the information source, such as a speaker, which are not intentional but nevertheless affect the destination, or listener. And the problem of semantic decoding must take this semantic noise into account. It is also possible to think of a treatment or adjustment of the original message that would make the sum of message meaning plus semantic noise equal to the desired total message meaning at the destination.

Another way in which the theory can be helpful in improving communication is suggested by the fact that error and confusion arise and fidelity decreases when, no matter how good the coding, one tries to crowd too much over a channel. A general theory at all levels will surely have to take into account not only the capacity of the channel but also (even the words are right!) the capacity of the audience. If you overcrowd the capacity of the audience, it is probably true, by direct analogy, that you do not fill the audience up and then waste only the remainder by spilling. More likely, and again by direct analogy, you force a general error and confusion.

The concept of information developed in this theory at first seems disappointing and bizarre—disappointing because it has nothing to do with meaning, and bizarre because it deals not with a single message but rather with the statistical character of a whole ensemble of messages, bizarre also because in these statistical terms the words information and uncertainty find themselves partners.

But we have seen upon further examination of the theory that this analysis has so penetratingly cleared the air that one is now perhaps for the first time ready for a real theory of meaning. An engineering communication theory is just like a very proper and discreet girl at the telegraph office accepting your telegram. She pays no attention to the meaning, whether it be sad or joyous or embarrassing. But she must be prepared to deal intelligently with all messages that come to her desk. This idea that a communication system ought to try to deal with all possible messages, and that the intelligent way to try is to base design on the statistical character of the source, is surely not without significance for communication in general. Language must be designed, or developed, with a view to the totality of things that man may wish to say; but not being able to accomplish everything, it should do as well as possible as often as possible. That is to say, it too should deal with its task statistically.

This study reveals facts about the statistical structure of the English language, as an example, which must seem significant to students of every phase of language and communication. It suggests, as a particularly promising lead, the application of probability theory to semantic studies. Especially pertinent is the powerful body of probability theory dealing with what mathematicians call the Markoff processes, whereby past events influence present probabilities, since this theory is specifically adapted to handle one of the most significant but difficult aspects of meaning, namely the influence of context. One has the vague feeling that information and meaning may prove to be something like a pair of canonically conjugate variables in quantum theory, that is, that information and meaning may be subject to some joint restriction that compels the sacrifice of one if you insist on having much of the other.

Or perhaps meaning may be shown to be analogous to one of the quantities on which the entropy or the thermodynamic ensemble depends. Here Eddington has another apt comment:

"Suppose that we were asked to arrange the following in two categories—*distance, mass, electric force, entropy, beauty, melody.*

"I think there are the strongest grounds for placing entropy alongside beauty and melody, and not with the first three. Entropy is only found when the parts are viewed in association, and it is by viewing or hearing the parts in association that beauty and melody are discerned. All three are features of arrangement. It is a pregnant thought that one of these three associates should be able to figure as a commonplace quantity of science. The reason why this stranger can pass itself off among the aborigines of the physical world is that it is able to speak their language, viz., the language of arithmetic."

One feels sure that Eddington would have been willing to include the word meaning along with beauty and melody; and one suspects he would have been thrilled to see, in this theory, that entropy not only speaks the language of arithmetic; it also speaks the language of language.

INFORMATION MACHINES
by Louis N. Ridenour

IF THE THERMOSTAT is a prime elementary example of the principle of automatic control, the computer is its most sophisticated expression. The thermostat and other simple control mechanisms, such as the automatic pilot and engine-governor, are specialized devices limited to a single function. An automatic pilot can control an airplane but would be helpless if faced with the problem of driving a car. Obviously for fully automatic control we must have mechanisms that simulate the generalized abilities of a human being, who can operate the damper on a furnace, drive a car or fly a plane, set a rheostat to control a voltage, work the throttle of an engine, and do many other things besides. The modern computer is the first machine to approach such general abilities.

Computer is really an inadequate name for these machines. They are called computers simply because computation is the only significant job that has so far been given to them. The name has somewhat obscured the fact that they are capable of much greater generality. When these machines are applied to automatic control, they will permit a vast extension of the control art—an extension from the use of rather simple specialized control mechanisms, which merely assist a human operator in doing a complicated task, to over-all controllers which will supervise a whole job. They will be able to do so more rapidly, more reliably, more cheaply and with just as much ingenuity as a human operator.

To describe its potentialities the computer needs a new name. Perhaps as good a name as any is "information machine." This term is intended to distinguish its function from that of a power machine, such as a loom. A loom performs the physical work of weaving a fabric; the information machine controls the pattern being woven. Its purpose is not the performance of work but the ordering and supervision of the way in which the work is done.

There are in current use two different kinds of information machine: the analogue computer and the digital computer. Several excellent popular articles have discussed the characteristics of these two types of computer; here we shall briefly recall their leading properties and then consider their respective possibilities as control mechanisms.

The analogue machine is just what its name implies: a physical analogy to the type of problem its designer wishes it to solve. It is modeled on the simple, specialized type of controller, such as a steam-engine governor. Information is supplied to the machine in terms of the value of some physical quantity—an electrical voltage or current, the degree of angular rotation of a shaft or the amount of compression of a spring. The machine transforms this physical quantity into another physical quantity in accordance with the rules of its construction. And since these rules have been chosen to simulate the rules governing the problem, the resulting physical quantity is the answer desired. If the analogue machine is being used as a control device, the final physical quantity is applied to exercise the desired control.

Consider, as an example, the flyball-governor pictured on page 5, whose purpose is to hold a steam engine to a constant speed. We notice, first, that information on the engine speed reaches the governor in the form of the speed of rotation of a shaft, while the output of the governor is expressed as the motion of a throttle which is closed or opened as the whirling balls rise or fall. Second, we notice that the relation between these two physical quantities is determined by the actual construction of the governor. The design of the controller has been dictated by its function.

In contrast to the analogue machine, a digital machine works by counting. Data on the problem must be supplied in the form of numbers; the machine processes this information in accordance with the rules of arithmetic or other formal logic, and expresses the final result in numerical form. There are two major consequences of this manner of working. First, input and output equipment must be designed to make an appropriate connection between the logical world of the digital machine and the physical world of the problem being solved or the process being controlled. Second, the problem to be solved must be formulated explicitly for the digital machine. In the case of the analogue machine, the problem is implicit in the construction of the machine itself; construction of a digital machine is determined not by any particular problem or class of problems but by the logical rules which the machine must follow in the solution of *any* problem presented.

Thus far the need for specialized input and output equipment, more than any other factor, has restricted the role of digital information machines to computing. In a computation, both the input and the output quantities are numbers, so the most rudimentary equipment will suffice to introduce the problem and register the result. There is no need (as there would be in a control application) to transform various physical quantities into numerical form before submitting them to the machine, or to transform the results of the calculation into a control action, such as moving a throttle. To use a digital information machine as a computer it is necessary only to provide (1) an input device such as a teletypewriter, which with the help of a human operator can translate printed numbers into signals intelligible to the machine, and (2) an output device such as a page-printer or electric typewriter, which can translate the signals generated by the machine into the printed numbers intelligible to men. Even this simple requirement, however, has not always been well met by the designers of information machines.

When a digital information machine is to be used as an instrument of control—and we can confidently expect that this will eventually be its major role—the design of input and output equipment becomes a more formidable task. While it is true that the structure of the machine itself depends on principles of logic rather than on the nature of its application, this is by no means true of the input and output elements. The input devices, or receptors, can use standard elements for receiving the program of instructions, but they must also receive data specifying the state of the particular process being controlled, and for this the detailed design will vary widely from one application to another. Similarly the effectors, which exercise the machine's control, must be designed in terms of the nature of the process or device being controlled.

In comparing digital and analogue machines as instruments for automatic control, we observe, first, that for simple control applications the analogue machine is almost always less elaborate than a digital machine would be. Even the most elementary digital machine requires an arithmetical (or logical) unit, a storage unit, a control unit, receptors and effectors. For simple problems, this array of equipment is wastefully elaborate. In contrast, an analogue machine need be no more complicated than the problem demands. A slide rule, for example, is a perfectly respectable information machine of the analogue type. The analogue machine's ability to do simple work by simple means explains its current predominance in the field of automatic control. The whole control art is so new and so little developed that most of the problems thus far tackled have been of a rather elementary nature.

As the control task becomes more complex, however, the analogue machine loses its advantage, and we begin to see a second fundamental difference between the two types of machine. The analogue machine is a physical analogy to the problem, and therefore the more complicated the problem, the more complicated the machine must be. If it is mechanical, longer and ever-longer trains of gears, ball-and-disk integrators and other devices must be connected together; if it is electrical, more and more amplifiers must be cascaded. In the mechanical case, the inevitable looseness in the gears and linkages, though tolerable in simple setups, will eventually add up to the point where the total "play" in the machine is bigger than the significant output quantities, and the device becomes useless. In the electrical case, the random electrical disturbances called "noise," which always occur in electrical circuits, will similarly build up until they overwhelm the desired signals. Since "noise" is far less obtrusive than "play," electrical analogue machines can be more complicated than their mechanical equivalents, but there is a limit. The great machine called Typhoon, built by the Radio Corporation of America for the simulation of flight performance in guided missiles, closely approaches that limit. It is perhaps the most complicated analogue device ever built, and very possibly the most complicated that it will ever be rewarding to build.

The digital machine, on the other hand, is entirely free of the hazards of "play" and "noise." There is no intrinsic limit to the complexity of the problem or process that a digital machine can handle or control. The switching system of our national telephone network, which enables any one of 50 million phones to be connected to any other, is a digital machine of almost unimaginable complexity.

The third important difference between analogue and digital machines is in their accuracy potential. The precision of the analogue machine is restricted by the accuracy with which physical quantities can be handled and measured. In practice, the best such a machine can achieve is an accuracy of about one part in 10,000; many give results accurate to only one or two parts in 100. For some applications this range of precision is adequate; for others it is not. On the other hand, a digital machine, which deals only with numbers, can be as precise as we wish to make it. To increase accuracy we need only increase the number of significant figures carried by the machine to represent each quantity being handled. Of course in a control operation the machine's over-all precision is limited by possible errors in translating physical quantities into numbers and vice versa, but this does not alter the fact that where high precision is required, a digital machine is usually preferable to the analogue type.

There is a fourth respect in which the two machines differ. An analogue machine works in what is called "real time." That is, it continuously offers a solution of the problem it is solving, and this solution is appropriate at every instant to all the input information which has so far entered the machine. If the machine is doing a mathematical problem, for example, it need not formulate explicitly the equations to be solved and then go through the steps of solving them, as a digital machine would have to do. The equations are inherent in the very structure of the machine, and it solves them by doing just what it was built to do. It can thus respond promptly to changing input data, and offer an up-to-date solution at every moment. This property of working in "real time" is very important in most problems of automatic control. An autopilot flying a plane must respond at once to an altitude change resulting from a gust of wind; the most precise information on how to adjust the flight controls will be worthless if it comes 30 seconds too late.

Since a digital machine works by formulating and solving an explicit logical model of the problem, it can work in "real time" only if the time it requires to obtain a solution, given new input data, is short compared with the period in which significant changes can take place in the system being controlled. Present-day digital machines can achieve this speed for many important problems—flight control of aircraft, for example—but they are not yet fast enough to handle all the "real-time" problems that we should like to turn over to them. It has been estimated that the fastest existing digital machines are some 20 times too slow to deal with the problem of simulating the complete flight performance of a high-speed guided missile—the problem that Typhoon was built to handle. As development proceeds, the operating rates of digital machines can be expected to increase rapidly.

We see, then, that both analogue and digital machines can be used for automatic control, and each has advantages in its own sphere. For simple applications in which no great precision is required, an analogue controller will usually be preferable. For complex problems, or problems in which high precision is required, a digital controller will be superior. Where "real-time" computations must be made, analogue machines are almost always used now, though digital machines are beginning to achieve speeds that fit them for this type of application.

All this refers to the present state of the art of automatic control. What can we guess about the developments to come?

The simple specialized analogue controllers already in use will surely be extended to wider application. But the most significant and exciting prospects reside in the digital machine. We can expect that it will soon open up a new dimension of control. The meaning of this prediction can be admirably illustrated in terms of the highly instrumented catalytic cracking plant which Eugene Ayres has described in a preceding chapter.

Mr. Ayres tells us of a plant in which there are some 150 different analogue controllers, each governing some aspect of the continuous process that the plant performs. Several hundred indicators on a central control panel offer the most detailed information on system performance. Many of these indicating instruments also provide continuous recordings. Manual controls which can override any automatic controller are present for use in emergency. The instruments and controls have been arranged on a flow diagram which simulates the organization of the plant and helps the human operator to find his way through the complexities of instrumentation. And the most important process-controls are adjusted manually according to the results of a periodic product analysis.

Clearly the human operator is still the master of this "automatic" plant. However elaborate the instrumentation, the readings of the instruments are still presented to men; however competent the automatic controllers, provision for human veto of their action is built into every one of them. Men are expected to meet emergencies, and to take control under "conditions of unstable equilibrium such as starting up or shutting down." The cracking plant is automatic only when the unexpected is not happening; in times of stress it falls back on human control, and its whole design is dictated by this necessity.

To this scheme there will soon be added end-point control—continuous adjustment of the main process-controls on the basis of a continuous product analysis within the system itself. This modification will improve performance, but it will leave the situation essentially as it was before: more routine responsibility will be given to machinery, but the human supervisor will still be vital to proper operation.

The digital information machine, employed as an instrument of supervisory automatic control, can change this picture radically. Since such a machine can be instructed to perform any set of logical operations, however complicated, it can be programed at the outset to react in emergencies precisely as would a well-instructed human operator—and it can react at least a thousand times faster. Further, the machine can be given a set of criteria for appraising the relative success of its various acts, and can be enabled to alter its own program of instructions in the light of experience on the job. Hence it will be capable of "learning" and of finding a better way to perform its operations than the one prescribed in the original instructions. And this universally adaptable

> In the completely automatic process, a computer will be used to coordinate the performance of the control points at each step in the process. The master control information will be supplied by automatic product analysis instruments such as those which now report to the quality control laboratory. The computer memory will be furnished with the data which will determine which control points should be corrected when there is a departure from standard in the product.

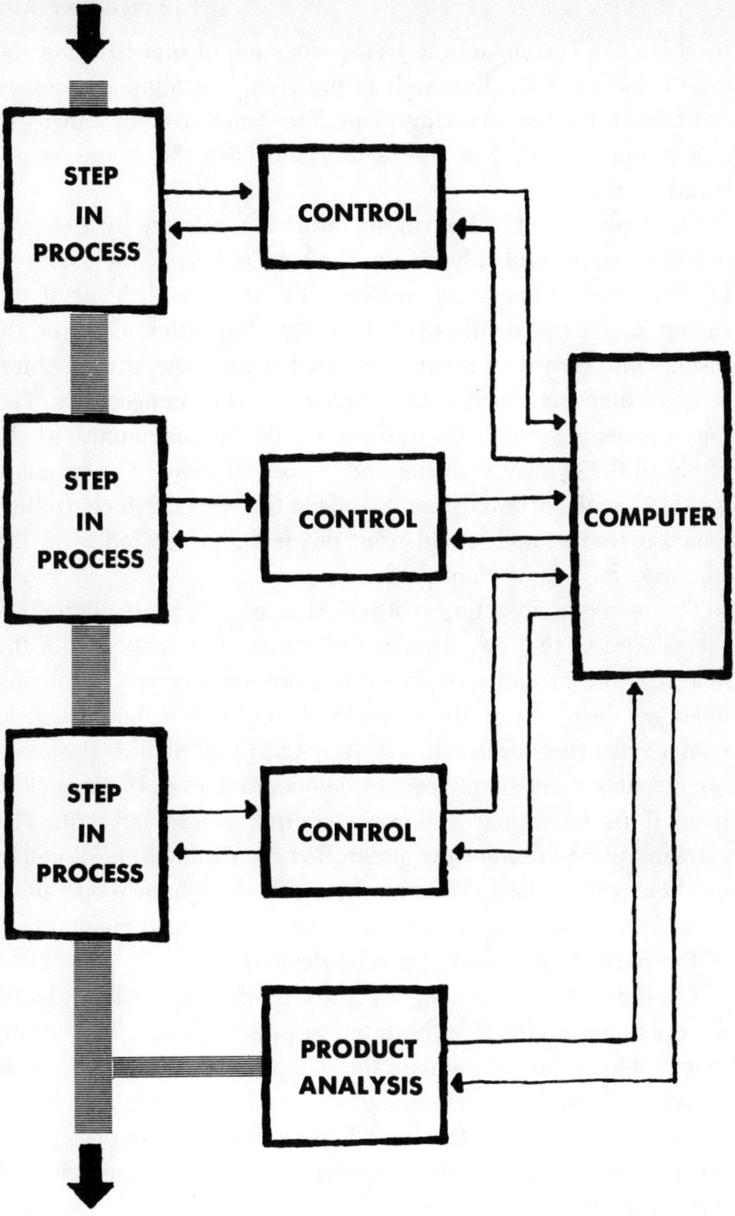
STEP
IN
PROCESS
CONTROL
STEP
IN
PROCESS
CONTROL
COMPUTER
STEP
IN
PROCESS
CONTROL
PRODUCT
ANALYSIS

machine can encompass the tremendous job of orchestrating the joint behavior of the hundreds of individual analogue controllers built into a modern cracking plant. The same machine can regulate the performance of the factory and keep the necessary accounting records.

The replacement of human operators in a refinery by a control machine would probably result in substantial economies, both in first cost and in operating cost per unit of product. Most of the saving in first cost would come from the elimination of the costly display and recording instruments that human operators require. In a machine-controlled plant, display would be unnecessary. The measurements vital to the process would be communicated directly to the control machine and processed there. The machine would issue the necessary commands to the specialized controllers which served it, and would print out in fully digested form the summary records of plant performance.

The saving in operating cost would come, not from eliminating the salaries of the few displaced operators, but simply from the fact that the machine could do a more efficient job. A human operator, even one of the greatest virtuosity, is a bottleneck in modern plant performance. Mr. Ayres has told us how the modern cracking plant simply cannot be operated, even by throngs of men, if its individual automatic controllers are left out. The cracking plant of tomorrow, controlled by a suitable information machine, will similarly be beyond the powers of human operators, even skillful ones equipped with all the control instrumentation—of the present variety—that can be devised.

The difficulty of designing a control room which will not baffle the operators is already substantial in present plants. This means that designers cannot increase the complexity of the plant, or its speed of operation, even though such changes might enhance efficiency. Removal of the limitations of human supervisors will open the way to vast design improvements. The information machine can remove them.

Some chemists think that a big new development in industrial chemistry lies just ahead, a development based on exploiting certain new types of reactions. These are fast reactions which take place within microseconds, reactions of gases flowing at velocities above the speed of sound, and reactions that will make it possible to capture valuable but fleeting intermediate products in a chemical system by preventing the system from reaching equilibrium. The enthusiasts say that the jet engine is the model of the chemical plant of the future. A supersonic chemical plant of the kind envisioned cannot be operated by men in white overalls reading carefully arranged gauges in an elaborate control room; the speed of nerve impulses from eye to brain to muscle is just too slow for that. Reactions occurring in microseconds must be controlled by machines that can respond in microseconds. Men will design these machines, build them and give them instructions, but men will never be able to compete with their performance.

If this last assertion seems outrageous, it is not more outrageous than it once was to assert that a man could design and build a derrick which would lift a load no man could ever budge. We are familiar with power machinery, and we take for granted its superiority to human muscles. We are not yet familiar with information machinery, and we are therefore not prepared to concede its superiority to the human nervous system. Nevertheless, a digital information machine can surpass human capabilities in any task that is governed by logical rules, no matter how complicated such rules may be.

Man's machines are beginning to operate at levels of speed, temperature, atomic radiation and complexity that make automatic control imperative. As an instrument of over-all automatic control the digital information machine has a great but as yet untouched potential. In the next few years this potential will begin to be realized, and the results are certain to be dramatic.

PART 4 **MACHINES AND MEN**

I. AN IMITATION OF LIFE
by W. Grey Walter

W. Grey Walter's interest in mechanical brains stems from his pioneering investigations of the living brain. Director of the Burden Neurological Institute in London, he has made fundamental contributions to the use of electroencephalography in investigations of the function of the brain and established some significant relationships between brain waves and the gross manifestations of emotion, personality and thought processes. He is a prolific writer and lecturer to the public in his field and one of the anchor men in the BBC Third Program ventures in popularization of science.

II. MAN VIEWED AS A MACHINE
by John G. Kemeny

At age twenty-eight, John G. Kemeny is professor of mathematics at Dartmouth College. He was born in Budapest, Hungary, came to the U. S. at the age of thirteen, graduated first in his class from George Washington High School in New York City and went on to Princeton University, whence he again graduated first in his class. His undergraduate education was interrupted by the U. S. Army; his service, however, consisted in assignment to the calculating machines at Los Alamos. After getting his doctoral degree in mathematics at Princeton, he had a last year of graduate study as Albert Einstein's assistant at the Institute for Advanced Study. His principal field of research is symbolic logic.

AN IMITATION OF LIFE
by W. Grey Walter

"When we were little . . . we went to school in the sea. The master was an old Turtle—we used to call him Tortoise."
"Why did you call him Tortoise if he wasn't one?" Alice asked.
"We called him Tortoise because he taught us," said the Mock Turtle angrily. "Really you are very dull!"

—Lewis Carroll,
Alice's Adventures in Wonderland

IN THE DARK ages before the invention of the electronic vacuum tube there were many legends of living statues and magic pictures. One of the commonest devices of sorcerers and witches was the model of an enemy which somehow embodied his soul, so that injury to the model would be reflected by suffering or death of the original. Even today it is not very uncommon to find in the cottages of European peasants wax statuettes of hated rivals stuck with pins and obscenely mutilated. One has only to recall the importance of graven images and holy pictures in many religions to realize how readily living and even divine properties are projected into inanimate objects by hopeful but bewildered men and women. Idolatry, witchcraft and other superstitions are so deeply rooted and widespread that it is possible even the most detached scientific activity may be psychologically equivalent to them; such activity may help to satisfy the desire for power, to assuage the fear of the unknown or to compensate for the flatness of everyday existence.

In any case there is an intense modern interest in machines that imitate life. The great difference between magic and the scientific imitation of life is that where the former is content to copy external appearance, the latter is concerned more with performance and behavior. Except in the comic strips the scientific robot does not look in the least like a living creature, though it may reproduce in great detail some of the complex functions which classical physiologists described as diagnostic of living processes. Some of the simpler of these functions can be duplicated by mechanical contrivances. But it was not until the electronic age that serious efforts were made to imitate and even to surpass the complex performance of the nervous system.

The fundamental unit of the nervous system is the nerve cell. In the human brain there are about 10,000 million such cells of various types, mostly concentrated in deep masses of "gray matter" or on the surface of the brain—the much-folded cerebral cortex. Between the cells run skeins of white matter, the interconnecting fibers. The unit of function is the nerve impulse, a miniature electro-chemical explosion that travels along the outside of a nerve fiber as a vortex ring of negative ions.

All the gradations of feeling and action of which we are capable are provided by variations in the frequency of nerve impulses and by the number of nerve cells stimulated. The brain cipher is even simpler than Morse code: it uses only dots, the number of which per second conveys all information. Communication engineers call this system "pulse-frequency modulation." It was "invented" by animals many millions of years ago, and it has advantages over other methods which are only just beginning to be applied. The engineers who have designed our great computing machines adopted this system without realizing that they were copying their own brains. (The popular term electronic brain is not so very fanciful.) In the language of these machines there are only two statements, "yes" and "no," and in their arithmetic only two numbers, 1 and 0. They surpass human capacity mainly in their great speed of action and in their ability to perform many interdependent computations at the same time, e.g., to solve simultaneous differential equations with hundreds of variables.

Magical though these machines may appear to the layman, their resemblance to living creatures is limited to certain details of their design. Above all they are in no sense free as most animals are free; rather they are parasites, depending upon their human hosts for nourishment and stimulation.

In a different category from computing machines are certain devices that have been made to imitate more closely the simpler types of living creatures, including their limitations (which in a computer would be serious faults) as well as their virtues. These less ambitious but perhaps more attractive mechanical creatures have evolved along two main lines. First there are stationary ones—sessile, the biologist would call them—which are rooted in a source of electric power and have very limited freedom. The prototype of these is the "homeostat" made by W. R. Ashby of Gloucester, England. It was created to study the mechanism whereby an animal adapts its total system to preserve its internal stability in spite of violent external changes.

The term "homeostasis" was coined by the Harvard University physiologist Walter B. Cannon to describe the many delicate biological mechanisms which detect slight changes of temperature or chemical state within the body and compensate for them by producing equal and opposite changes. Communication engineers have rediscovered this important expedient in their grapplings with the problems of circuits and computers. Their "negative feedback," which neutralizes departures from the desired state in a communication system, corresponds exactly to the Cannon notion. In animals most of what is called reflex activity has the property of feedback.

In Ashby's homeostat there are a number of electronic circuits similar to the reflex arcs in the spinal cord of an animal. These are so combined with a number of radio tubes and relays that out of many thousands of possible connections the machine will automatically find one that leads to a condition of dynamic internal stability. That is, after several trials and errors the instrument establishes connections which tend to neutralize any change that the experimenter tries to impose from outside. It is a curious fact that although the machine is man-made, the experimenter finds it impossible to tell at any moment exactly what the machine's circuit is without "killing" it and dissecting out the "nervous system"; that is, switching off the current and tracing out the wires to the relays. Nevertheless the homeostat does not behave very like an active animal—it is more like a sleeping creature which when disturbed stirs and finds a comfortable position.

Another branch of electromechanical evolution is represented by the little machines we have made in Bristol. We have given them the mock-biological name *Machina speculatrix,* because they illustrate particularly the exploratory, speculative behavior that is so characteristic of most animals. The machine on which we have chiefly concentrated is a small creature with a smooth shell and a protruding neck carrying a single eye which scans the surroundings for light stimuli; because of its general appearance we call the genus "Testudo," or tortoise. The Adam and Eve of this line are nicknamed Elmer and Elsie, after the initials of the terms describing them—ELectro MEchanical Robots, Light-Sensitive, with Internal and External stability. Instead of the 10,000 million cells of our brains, Elmer and Elsie contain but two functional elements: two miniature radio tubes, two sense organs, one for light and the other for touch, and two effectors or motors, one for crawling and the other for steering. Their power is supplied by a miniature hearing-aid B battery and a miniature six-volt storage battery, which provides both A and C current for the tubes and the current for the motors.

The number of components in the device was deliberately restricted to two in order to discover what degree of complexity of behavior and independence could be achieved with the smallest number of elements connected in a system providing the greatest number of possible interconnections. From the theoretical standpoint two elements equivalent to circuits in the nervous system can exist in six modes; if one is called A and the other B, we can distinguish A, B, A + B, A→B, B→A and A⇌B as possible dynamic forms. To indicate the variety of behavior possible for even so simple a system as this, one need only mention that six elements would be more than enough to form a system which would provide a new pattern every tenth of a second for 280 years—four times the human lifetime of 70 years! It is unlikely that the number of perceptible functional elements in the human brain is anything like the total number of nerve cells; it is more likely to be of the order of 1,000. But even if it were only 10, this number of elements could provide enough variety for a lifetime of experience for all the men who ever lived or will be born if mankind survives a thousand million years.

So a two-element synthetic animal is enough to start with. The strange richness provided by this particular sort of permutation introduces right away one of the aspects of animal behavior—and human psychology—which *M. speculatrix* is designed to illustrate: the uncertainty, randomness, free will or independence so strikingly absent in most well-designed machines. The fact that only a few richly interconnected elements can provide practically infinite modes of existence suggests that there is no logical or experimental necessity to invoke more than *number* to account for our subjective conviction of freedom of will and our objective awareness of personality in our fellow men.

The behavior of Elmer and Elsie is in fact remarkably unpredictable. The photocell, or "eye," is linked with the steering mechanism. In the absence of an adequate light-stimulus Elmer (or Elsie) explores continuously, and at the same time the motor drives it forward in a crawling motion. The two motions combined give the creature a cycloidal gait, while the photocell "looks" in every direction in turn. This process of scanning and its synchronization with the steering device may be analogous to the mechanism whereby the electrical pulse of the brain known as the alpha rhythm sweeps over the visual brain areas and at the same time releases or blocks impulses destined for the muscles of the body. In both cases the function is primarily one of economy, just as in a television system the scanning of the image permits transmission of hundreds of thousands of point-details on one channel instead of on as many channels.

The effect of this arrangement on Elmer is that in the dark it explores in a very thorough manner a considerable area, remaining alert to the possibility of light and avoiding obstacles that it cannot surmount or push aside. When the photocell sees a light, the resultant signal is amplified by both tubes in the amplifier. If the light is very weak, only a *change* of illumination is transmitted as an effective signal. A slightly stronger signal is amplified without loss of its absolute level. In either case the effect is to halt the steering mechanism so that the machine moves toward the light source or maneuvers so that it can approach the light with the least difficulty. This behavior is of course analogous to the reflex behavior known as "positive tropism," such as is exhibited by a moth flying into a candle. But Elmer does not blunder into the light, for when the brilliance exceeds a certain value—that of a flashlight about six inches away—the signal becomes strong enough to operate a relay in the first tube, which has the reverse effect from the second one. Now the steering mechanism is turned on again at double speed, so the creature abruptly sheers away and seeks a more gentle climate. If there is a single light source, the machine circles around it in a complex path of advance and withdrawal; if there is another light farther away, the machine will visit first one and then the other and will continually stroll back and forth between the two. In this way it neatly solves the dilemma of Buridan's ass, which the scholastic philosophers said would die of starvation between two barrels of hay if it did not possess a transcendental free will.

For Elmer hay is represented, of course, by the electricity it needs to recharge its batteries. Within the hutch where it normally lives is a battery charger and a 20-watt lamp. When the creature's batteries are well charged, it is attracted to this light from afar, but at the threshold the brilliance is great enough to act as a repellent, so the model wanders off for further exploration. When the batteries start to run down, the first effect is to enhance the sensitivity of the amplifier so that the attraction of the light is felt from even further away. But soon the level of sensitivity falls and then, if the machine is fortunate and finds itself at the entrance to its kennel, it will be attracted right home, for the light no longer seems so dazzling. Once well in, it can make contact with the charger. The moment current flows in the circuit between the charger and the batteries the creature's own nervous system and motors are automatically disconnected; charging continues until the battery voltage has risen to its maximum. Then the internal circuits are automatically reconnected and the little creature, repelled now by the light which before the feast had been so irresistible, circles away for further adventures.

Inevitably in its peripatetic existence *M. speculatrix* encounters many obstacles. These it cannot "see," because it has no vestige of pattern vision, though it will avoid an obstacle that casts a shadow when it is approaching a light. The creature is equipped, however, with a device that enables it to get around obstacles. Its shell is suspended on a single rubber mounting and has sufficient flexibility to move and close a ring contact. This contact converts the two-stage amplifier into a multivibrator. The oscillations so generated rhythmically open and close the relays that control the full power to the motors for steering and crawling. At the same time the amplifier is prevented from transmitting the signals picked up by the photocell. Accordingly when the creature makes contact with an obstacle, whether in its speculative or tropistic mode, all stimuli are ignored and its gait is transformed into a succession of butts, withdrawals and sidesteps until the interference is either pushed aside or circumvented. The oscillations persist for about a second after the obstacle has been left behind; during this short memory of frustration Elmer darts off and gives the danger area a wide berth.

When the models were first made, a small light was connected in the steering-motor circuit to act as an indicator showing when the motor was turned off and on. It was soon found that this light endowed the machines with a new mode of behavior. When the photocell sees the indicator light in a mirror or reflected from a white surface, the model flickers and jigs at its reflection in a manner so specific that were it an animal a biologist would be justified in attributing to it a capacity for self-recognition. The reason for the flicker is that the vision of the light results in the indicator light being switched off, and darkness in turn switches it on again, so an oscillation of the light is set up.

Two creatures of this type meeting face to face are affected in a similar but again distinctive manner. Each, attracted by the light the other carries, extinguishes its own source of attraction, so the two systems become involved in a mutual oscillation, leading finally to a stately retreat. When the encounter is from the side or from behind, each regards the other merely as an obstacle; when both are attracted by the same light, their jostling as they approach the light eliminates the possibility of either reaching its goal. When one machine casually interferes with another while the latter is seriously seeking its charging light, a dog-in-the-manger situation develops which results in the more needy one expiring from exhaustion within sight of succor.

These machines are perhaps the simplest that can be said to resemble animals. Crude though they are, they give an eerie impression of purposefulness, independence and spontaneity. More complex models that we are now constructing have memory circuits in which associations are stored as electric oscillations, so the creatures can learn simple tricks, forget them slowly and relearn more quickly. This compact, plastic and easily accessible form of short-term memory may be very similar to the way in which the brain establishes the simpler and more evanescent conditioned reflexes.

One intriguing effect in these higher forms of synthetic life is that as soon as two receptors and a learning circuit are provided, the possibility of a conflict neurosis immediately appears. In difficult situations the creature sulks or becomes wildly agitated and can be cured only by rest or shock—the two favorite stratagems of the psychiatrist. It appears that it would even be technically feasible to build processes of self-repair and of reproduction into these machines.

Perhaps we flatter ourselves in thinking that man is the pinnacle of an estimable creation. Yet as our imitation of life becomes more faithful our veneration of its marvelous processes will not necessarily become less sincere.

MAN VIEWED AS A MACHINE
by John G. Kemeny

Is MAN NO more than a machine? The question is often debated these days, usually with more vigor than precision. More than most arguments, this one tends to bog down in definition troubles. What is a machine? And what do we mean by "no more than"? If we define "machine" broadly enough, everything is a machine; and if by "more than" we mean that we are human, then machines are clearly less than we are.

In this article we shall frame the question more modestly. Let us ask: What could a machine do as well or better than a man, now or in the future? We shall not concern ourselves with whether a machine could write sonnets or fall in love. Nor shall we waste time laboring the obvious fact that when it comes to muscle, machines are far superior to men. What concerns us here is man as a brain-machine. John von Neumann, the mathematician and designer of computers, not long ago made a detailed comparison of human and mechanical brains in a series of lectures at Princeton University. Much of what follows is based on that discussion.

In economy of energy the human brain certainly is still far ahead of all its mechanical rivals. The entire brain with its many billions of cells functions on less than 100 watts. Even with the most efficient present substitute for a brain cell—the transistor—a machine containing as many cells as the brain would need about 100 million watts. We are ahead by a factor of at least a million. But Von Neumann has calculated that in theory cells could be 10 billion times more efficient in the use of energy than the brain cells actually are. Thus there seems to be no technical reason why mechanical brains should not become more efficient energy-users than their human cousins. After all, just recently by inventing the transistor, which requires only about a hundredth of a watt, we have improved the efficiency of our machines by a factor of 100; in view of this the factor of a million should not frighten us.

While we are still ahead in the use of energy, we are certainly far behind in speed. Whereas a nerve cannot be used more than 100 times a second, a vacuum tube can easily be turned on and off a million times a second. It could be made to work even faster, but this would not contribute much to speeding up the mechanical brain at the moment. No machine is faster than its slowest part, so we must evaluate various components of the machine.

In a calculating machine four different problems confront the designer: the actual computations, the "logical control," the memory and the feeding of information to the machine and getting answers out. Speed of computation, a bottleneck in mechanical computers such as the desk calculator, has been taken care of by the vacuum tube. The next bottleneck was the logical control—the system for telling the machine what to do next after each step. The early IBM punch-card machine took this function out of the hands of a human operator by using a wiring setup on a central board which commanded the sequence of operations. This is perfectly all right as long as the machine has to perform only one type of operation. But if the sequence has to be changed frequently, the wiring of the board becomes very clumsy indeed. To improve speed the machine must be given an internal logical control. Perhaps the greatest step forward on this problem has been accomplished by MANIAC, built at the Institute for Advanced Study in Princeton. This machine can change instructions as quickly as it completes calculations, so that it can operate as fast as its vacuum tubes will allow.

That still leaves the problems of speeding up the memory and the input and output of information. The two problems are closely related. The larger the memory, the less often the operator has to feed the machine information. But the very fact that the machine performs large numbers of computations between instructions clogs its memory and slows it down. This is because an accumulation of rounding errors makes it necessary to carry out all figures in a calculation to a great number of digits. In each computation the machine necessarily rounds off the last digit; in succeeding operations the digit becomes less and less precise. If the computations are continued, the next-to-last digit begins to be affected, and so on. It can be shown that after 100 computations the last digit is worthless; after 10,000 the last two digits; after 1,000,000, the last three. In the large new computers an answer might easily contain four worthless figures. Hence to insure accuracy the machine must carry more digits than are actually significant; it is not uncommon to carry from eight to 12 digits for each number throughout the calculation. When the machine operates on the binary system of numbers, instead of the decimal system, the situation is even worse, for it takes about three times as many digits to express a number in the binary scale.

MANIAC uses up to 40 binary digits to express a number. Due to the necessity for carrying this large number of digits, even MANIAC's celebrated memory can hold no more than about 1,000 numbers. It has an "external memory," in the form of a magnetic tape and magnetic drums, in which it can store more information, but reading from the tape or drums is a much slower operation than doing electrical computations.

In spite of the present limitations, the machines already are ahead of the human brain in speed by a factor of at least 10,000 —usually a great deal more than 10,000. They are most impressive on tasks such as arise in astronomy or ballistics. It would be child's play for MANIAC to figure out the position of the planets for the next million years.

Still we are left with the feeling that there are many things we can do that a machine cannot do. The brain has more than 10 billion cells, while a computer has only a few tens of thousands of parts. Even with transistors, which overcome the cost and space problems, the difficulty of construction will hardly allow more than a million parts to a machine. So we can safely say that the human brain for a long time to come will be about 10,000 times more complex than the most complicated machine. And it is well known that an increase of parts by a factor of 10 can bring about differences in kind. For example, if we have a unit that can do addition and multiplication, by combining a few such units with a logical control mechanism we can do subtraction, division, raising to powers, interpolation and many other operations qualitatively different from the original.

Part of man's superior complexity is his remarkable memory. How does MANIAC's memory compare with it? For simplicity's sake let us measure the information a memory may hold in "bits" (for binary digits). A vacuum tube can hold one digit of a binary number (the digit is 1 if the tube is on, 0 if it is off). In vacuum-tube language it takes 1,500 bits to express the multiplication table. Now MANIAC's memory holds about 40,000 bits, not in 40,000 separate tubes but as spots on 40 special picture tubes, each of which can hold about 1,000 spots (light or dark). Estimates as to how much the human memory holds vary widely, but we certainly can say conservatively that the brain can remember at least 1,000 items as complex as the multiplication table (1.5 million bits), and a reasonable guess is that its capacity is closer to 100 million bits—which amounts to acquiring one bit per 20 seconds throughout life. So our memory exceeds that of MANIAC by a factor of 1,000 at least.

Is the difference just a matter of complexity? No, the fact is that machines have not yet imitated the human brain's method of storing and recovering information. For instance, if we tried to increase MANIAC's memory by any considerable amount, we would soon find it almost impossible to extract information. We would have to use a complex system of coding to enable the machine to hunt up a given item of information, and this coding would load down the memory further and make the logical control more complex. Only when we acquire a better understanding of the brain's amazing ability to call forth information will we be able to give a machine anything more than a limited memory.

Let us now consider the inevitable question: Can a machine "think"? We start with a simple model of the nervous system such as has been constructed by Walter Pitts and Warren S. McCulloch of the Massachusetts Institute of Technology. Its basic unit is the neuron—a cell that can be made to emit pulses of energy. The firing of one neuron may activate the next or it may inhibit it. The neurons are assumed to work in cycles. This corresponds to our knowledge that after firing a neuron must be inactive for a period. To simplify the model it is assumed that the various neurons' cycles are synchronized, i.e., all the neurons active during a given period fire at the same time. For a given neuron to fire in a given cycle two conditions must be satisfied: in the

Three simple circuits for a brain machine are illustrated in these diagrams. The panel at left provides the key to the symbols. The panel second from left illustrates the basic cell, which fires when it is activated (top) but does not fire if it is inhibited at the same

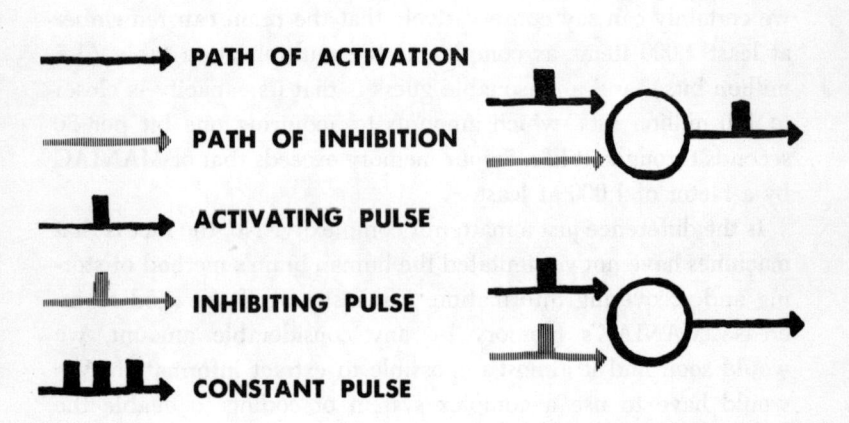

⟶ **PATH OF ACTIVATION**

⇢ **PATH OF INHIBITION**

⟶ **ACTIVATING PULSE**

⇢ **INHIBITING PULSE**

⟶ **CONSTANT PULSE**

previous cycle it must have been (1) activated and (2) not inhibited. If, for example, a neuron has two others terminating in it of which one activates and one inhibits, and if the former fires in a given cycle and the latter does not, then the neuron will fire in the following cycle. Otherwise it will be inactive for a cycle.

Out of this basic pattern we can build the most complex logical machine. We can have a combination that will fire if a connected neuron did not fire (representing "not") or one that will fire if at least one of the two incoming neurons fired (representing "or") or one that will fire only if both incoming neurons fired (representing "and"). Combining these, we can imitate many logical operations of the brain.

We can also construct a very primitive memory: e.g., a system that will "remember" that it has been activated until it is instructed to "forget" it. But if it is to remember anything at all complex, it must have an unthinkably large number of neurons—

time (bottom). The "or" circuit, in the next panel, has two paths of activation and fires when a pulse arrives on one path "or" the other. The "not" circuit, at right, is fired by a constant stream of activating pulses.

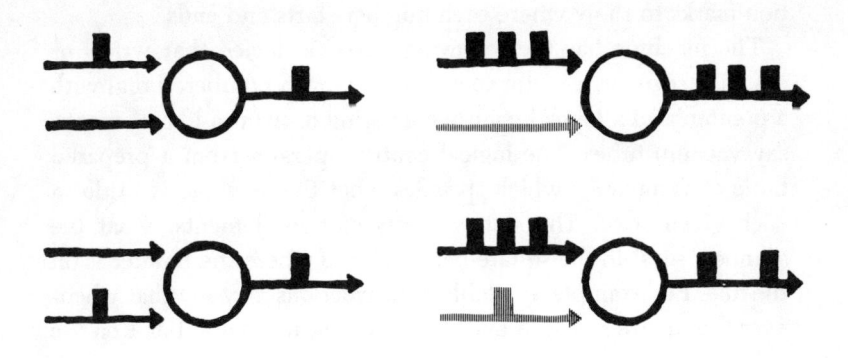

another illustration of the fact that human memory acts on different principles from a machine.

If we were to stop here, we might conclude that practical limitations of memory and complexity must forever restrict the cleverness or versatility of any machine. But we have not yet plumbed the full possibilities. The late A. M. Turing of England showed, by a brilliant analysis, that by combining a certain few simple operations in sufficient number a machine could perform feats of amazing complexity. Turing's machines may be clumsy and slow, but they present the clearest picture of what a machine can do.

A Turing machine can be thought of as a mechanical calculator which literally works with pencil and paper. The paper it uses is a long tape divided into successive squares, and it operates on one square at a time. As it confronts a particular square it can do one of six things: (1) write down the letter X; (2) write down the digit 1; (3) erase either of these marks if it is already in the square; (4) move the tape one square to the left; (5) move the tape one square to the right; (6) stop.

Essentially this machine is a number writer. It writes its numbers in the simplest possible form, as a string of units. This is even simpler than the binary system. In the binary system the number 35, for example, is written 100011. In a Turing machine it is a string of 1's in 35 successive squares. The X's are merely punctuation marks to show where each number starts and ends.

The machine has the following parts: a device that writes or erases, a scanner, a motor to move the tape, a numbered dial with a pointer, and a logical control consisting of neuron-like elements, say vacuum tubes. The logical control operates from a prepared table of commands which specifies what the machine is to do in each given state. The state consists of two elements: what the scanner "sees" in the square before it, and where the pointer is on the dial. For example, the table of instructions may say that whenever the square has an X and the pointer is at the number 1 on the dial, the machine is to erase the X, and move the pointer to the number 2 on the dial. As the machine proceeds from step to step, the logical control gives it such commands, the command in each case depending both on the position of the dial and on what the scanner sees in the square confronting it. Observe that the dial functions as a primitive "memory," in the sense that its position at any stage is a consequence of what the scanner saw and where the pointer stood at the step immediately preceding. It carries over the machine's experience from step to step.

Turing's machine thus consists of a tape with X's and 1's in some of its squares, a dial-memory with a certain number of positions, and a logical control which instructs the machine what to do, according to what it sees and what its memory says. A very simple version of the machine might be imagined as having a dial with only six positions. Since the scanner may see one of three things in a square—blank, 1 or X—the machine has 18 possible states, and the logical control has a command for each case. This machine is designed to perform a single task: it can add two numbers—any two numbers. Suppose it is to add 2 and 3. The numbers are written as strings of 1's with X's at the ends. Say we start with the dial at position 1 and the scanner looking at the second digit of the number 3. The instructions in the table say that when it is in this state the machine is to move the tape one square to the right and keep the dial at position 1. This operation brings the square to the left, containing another digit 1, under the scanner. Again the instructions are the same: "Move the tape one square to the right and keep the dial at position 1." Now the scanner sees an X. The instructions, with the dial at position 1, are: "Erase (the X) and move the dial to position 2." The machine now confronts a blank square. The command becomes: "Move the tape one square to the right and keep the dial at position 2." In this manner the machine will eventually write two digit 1's next to the three at the right and end with the answer 5—a row of five digits enclosed by X's. When it finishes, an exclamation point signifies that it is to stop. The reader is advised to try adding two other numbers in the same fashion.

This surely is a cumbersome method of adding. However, the machine becomes more impressive when it is expanded so that it can solve a problem such as the following: "Multiply the number you are looking at by two and take the cube root of the answer if the fifth number to the left is less than 150." By adding positions to the dial and enlarging the table of instructions we can endow such a machine with the ability to carry out the most complex tasks, though each operational step is very simple. The Turing machine in fact resembles a model of the human nervous system, which can be thought of as having a dial with very many positions and combining many simple acts to accomplish the enormous number of tasks a human being is capable of.

Turing gave his machines an infinite memory. Of course the dial can have only a finite number of positions, but he allowed the machine a tape infinite in length, endless in both directions. Actually the tape does not have to be infinite—just long enough for the task. We may provide for all emergencies by allowing the machine to ask for more tape if it needs it. The human memory is infinite in the same sense: we can always make more paper to make notes on.

If we allow the unlimited tape, the Turing idea astounds us further with a universal machine. Not only can we build a machine for each task, but we can design a single machine that is versatile enough to accomplish all these tasks! We must try to understand how this is done, because it will give us the key to our whole problem.

The secret of the universal machine is that it can imitate. Suppose we build a highly complex machine for a difficult task. If we then supply the universal machine with a description of the task and of our special machine, it will figure out how to perform the task. It proceeds very simply, deducing from what it knows about our machine just what it would do at each step. Of course this slows the universal machine down considerably. Between any two steps it must carry out a long argument to analyze what our machine would do. But we care only about its ability to succeed, not its speed. There is no doubt about it: anything any logical machine can do can be done by this single mechanism.

The key question is: How do you describe a complex machine in terms that a relatively simple machine can understand? The answer is that you devise a simple code which can describe any machine (or at least any Turing machine), and that you design the universal machine so that it will be able to understand this code. To understand a Turing machine we need only know its table of commands, so it suffices to have a simple code for tables of commands.

The universal machine is remarkably human. It starts with very limited abilities, and it learns more and more by imitation and by absorbing information from the outside. We feel that the potentialities of the human brain are inexhaustible. But would this be the case if we were unable to communicate with the world around us? A man robbed of his five senses is comparable to a Turing machine with a fixed tape, but a normal human being is like the universal machine. Given enough time, he can learn to do anything.

But some readers will feel we have given in too soon to Turing's persuasive argument. After all a human being must step in and give the universal machine the code number. If we allow that, why not give the machine the answer in the first place? Turing's reply would have been that the universal machine does not need a man to encode the table; it can be designed to do its own coding, just as it can be designed to decode.

So we grant this amazing machine its universal status. And

although its table of logical control has only a few thousand entries, it seems to be able to do essentially all the problem-solving tasks that we can. Of course it might take a billion years to do something we can do in an hour. The "outside world" from which it can learn is much more restricted than ours, being limited to Turing machines. But may not all this be just a difference of degree? Are we, as rational beings, basically different from universal Turing machines?

The usual answer is that whatever else machines can do, it still takes a man to build the machine. Who would dare to say that a machine can reproduce itself and make other machines?

Von Neumann would. As a matter of fact, he has blueprinted just such a machine.

What do we mean by reproduction? If we mean the creation of an object like the original out of nothing, then no machine can reproduce, but neither can a human being. If reproduction is not to violate the conservation of energy principle, building materials must be available. The characteristic feature of the reproduction of life is that the living organism can create a new organism like itself out of inert matter surrounding it.

If we agree that machines are not alive, and if we insist that the creation of life is an essential feature of reproduction, then we have begged the question: A machine cannot reproduce. So we must reformulate the problem in a way that won't make machine reproduction logically impossible. We must omit the word "living." We shall ask that the machine create a new organism like itself out of simple parts contained in the environment.

Human beings find the raw material in the form of food; that is, quite highly organized chemicals. Thus we cannot even say that we produce order out of complete disorder, but rather we transform more simply organized matter into complex matter. We must accordingly assume that the machine is surrounded with pieces of matter, simpler than any part of the machine. The hypothetical parts list would be rolls of tape, pencils, erasers, vacuum tubes, dials, photoelectric cells, motors, shafts, wire, batteries and so on. We must endow the machine with the ability to transform pieces of matter into these parts and to organize them into a new machine.

Von Neumann simplified the problem by making a number of reasonable assumptions. First of all he realized that it is inessential for the machine to be able to move around. Rather, he has the mechanism sending out impulses which organize the surroundings by remote control. Secondly, he assumed that space is divided into cubical cells, and that each part of the machine and each piece of raw material occupies just one cell. Thirdly, he assumed that the processes are quantized not only in space but in time; that is, we have cycles during which all action takes place. It is not even necessary to have three dimensions: a two-dimensional lattice will serve as well as the network of cubes.

Our space will be a very large (in principle infinite) sheet, divided into squares. A machine occupies a connected area consisting of a large number of squares. Since each square represents a part of the machine, the number of squares occupied is a measure of the complexity of the machine. The machine is surrounded by inert cells, which it has to organize. To make this possible the machine must be a combination of a brain and a brawn machine, since it not only organizes but also transforms matter. Accordingly the Von Neumann machine has three kinds of parts. It has neurons similar to those discussed in the model of the nervous system. These provide the logical control. Then it has transmission cells, which carry messages from the control centers. They have an opening through which they can receive impulses, and an output through which the impulse is passed on a cycle later. A string of transmission cells, properly adjoined, forms a channel through which messages can be sent. In addition the machine has muscles. These cells can change the surrounding cells, building them up from less highly organized to more complex cells or breaking them down. They bring about changes analogous to those produced by a combination of muscular and chemical action in the human body. Their primary use is, of course, the changing of an inert cell into a machine part.

As in the nervous system, the operation proceeds by steps: the state of every cell is determined by its state and the state of its neighbors a cycle earlier. The neurons and transmission cells are either quiescent or they can send out an impulse if properly stimulated. The muscle cells receive commands from the neurons through the transmission cells, and react either by "killing" some undesired part (i.e., making it inert) or by transforming some inert cell in the environment to a machine part of a specific kind. So far the machine is similar in structure to a higher animal. Its neurons form the central nervous system; the transmission cells establish contact with various organs; the organs perform their designated tasks upon receiving a command.

The instructions may be very long. Hence they must in a sense be external. Von Neumann's machine has a tail containing the blueprint of what it is to build. This tail is a very long strip containing coded instructions. The basic box performs two types of functions: it follows instructions from its tail, and it is able to copy the tail. Suppose the tail contains a coded description of the basic box. Then the box will, following instructions, build another box like itself. When it is finished, it proceeds to copy its own tail, attaching it to the new box. And so it reproduces itself.

The secret of the machine is that it does not try to copy itself. Von Neumann designed a machine that can build any machine from a description of it, and hence can build one like itself. Then it is an easy matter to copy the large but simple tail containing the instructions and attach it to the offspring. Thereafter the new machine can go on producing more and more machines until all the raw material is used up or until the machines get into conflict with each other—imitating even in this their human designers.

It is amazing to see how few parts such a machine needs to have. Von Neumann's blueprints call for a basic box of 80 by 400 squares, plus a tail 150,000 squares long. The basic box has the three kinds of parts described—neurons, transmission cells and muscle cells. The three types of cells differ only as to their state of excitation and the way in which they are connected. The tail is even simpler: it has cells, which are either "on" or "off," holding a code. So we have about 200,000 cells, most of which are of the simplest possible kind, and of which only a negligible fraction is even as complex as the logical control neuron. No matter how we measure complexity, this is vastly simpler than a human being, and yet the machine is self-reproducing.

Pressing the analogy between the machine and the human organism, we might compare the tail to the set of chromosomes. Our machine always copies its tail for the new machine, just as each daughter cell in the body copies the chromosomes of its parent. It is most significant that while the chromosomes take up a minute part of the body, the tail is larger than the entire basic box in the machine. This indicates that the coding of traits by chromosomes is amazingly efficient and compact. But in all fairness we must point out that the chromosomes serve a lesser role than the tail. The tail contains a complete description of the basic box, while the chromosome description is incomplete: the offspring only resembles the parent; it is not an exact duplicate. It would be most interesting to try to continue Von Neumann's pioneer work by designing a machine that could take an incomplete description and build a reasonable likeness of itself.

Could such machines go through an evolutionary process? One might design the tails in such a way that in every cycle a small number of random changes occurred (e.g., changing an "on" to an "off" in the code or vice versa). These would be like mutations; if the machine could still produce offspring, it would pass the changes on. One could further arrange to limit the supply of raw material, so that the machines would have to compete for *Lebensraum,* even to the extent of killing one another.

Of course none of the machines described in this article has actually been built, so far as I know, but they are all buildable. We have considered systematically what man can do, and how much of this a machine can duplicate. We have found that the brain's superiority rests on the greater complexity of the human nervous system and on the greater efficiency of the human memory. But is this an essential difference, or is it only a matter of degree that can be overcome with the progress of technology? This article attempted to show that there is no conclusive evidence for an essential gap between man and a machine. For every human activity we can conceive of a mechanical counterpart.

Naturally we still have not answered the question whether man is more than a machine. The reader will have to answer that question for himself.

BIBLIOGRAPHY

READERS interested in further reading on the topics covered in this book may find the list below helpful. It is *not* a bibliography of source material. The books chosen are for the most part addressed to the general reader; they include also some of the more accessible textbooks and survey volumes. The list is by no means exhaustive. Nor does it embrace the full range of interest of this book, since much of the work reported here is not yet represented in the pages of any other book. (The date given in italics under each chapter title is the date of its original publication in SCIENTIFIC AMERICAN.)

SELF-REGULATION
September 1952

The Human Use of Human Beings. Norbert Wiener. Houghton Mifflin Company, 1950.

FEEDBACK
September 1952

Fundamentals of Automatic Control. G. H. Farrington. John Wiley & Sons, Inc., 1951.
An Introduction to the Theory of Control in Mechanical Engineering. R. H. Macmillan. Cambridge University Press, 1951.
Servomechanism Fundamentals. Henri Lauer, Robert Lesnick and Leslie E. Matson. McGraw-Hill Book Company, Inc., 1947.
Fundamental Theory of Servomechanisms. LeRoy A. MacColl. D. Van Nostrand Company, Inc., 1945.
Theory of Servomechanisms. H. M. James, N. B. Nichols and R. S. Phillips. McGraw-Hill Book Company, Inc., 1947.
Automatic Feedback Control. W. R. Ahrendt and J. F. Taplin. McGraw-Hill Book Company, Inc., 1951.

CONTROL SYSTEMS
September 1952

Automatic and Manual Control. A. Tustin. Butterworths Scientific Publications, London, 1952.
Principles of Servomechanisms. G. S. Brown and D. P. Campbell. John Wiley & Sons, Inc., 1948.

BIBLIOGRAPHY

An Introduction to Servomechanisms. A. Porter. John Wiley & Sons, Inc., 1950.

Servomechanism and Regulating-System Design, Vol. I. Harold Chestnut and Robert W. Mayer. General Electric Series, John Wiley & Sons, Inc., 1951.

AN AUTOMATIC CHEMICAL PLANT
September 1952

Industrial Instruments for Measurement and Control. Thomas J. Rhodes. McGraw-Hill Book Company, Inc., 1941.

Industrial Instrumentation. D. P. Eckman. John Wiley & Sons, Inc., 1950.

AN AUTOMATIC MACHINE TOOL
September 1952

English and American Tool Builders. Joseph Wickham Roe. Yale University Press, 1916.

THE AUTOMATIC OFFICE
January 1954

Automation. John Diebold. D. Van Nostrand Company, Inc., 1952.

"Electronics and the Banks." M. S. Goldring in *The Banker*; March, April, May, 1953.

Faster Than Thought. B. V. Bowden. Pitman Publishing Corporation, 1953.

THE ECONOMIC IMPACT
September 1952

A History of Mechanical Invention. Abbott P. Usher. McGraw-Hill Book Company, Inc., 1929.

Mechanization in Industry. Harry Jerome. Publication No. 27, National Bureau of Economic Research, 1934.

Trends in Output and Employment. George J. Stigler. National Bureau of Economic Research, 1947.

WHAT IS INFORMATION?
September 1952

Bibliography in an Age of Science. Louis N. Ridenour, Ralph R. Shaw and Albert C. Hill. University of Illinois Press, 1951.

THE MATHEMATICS OF INFORMATION
July 1949

A Mathematical Theory of Communication. Claude E. Shannon and Warren Weaver. University of Illinois Press, 1949.

INFORMATION MACHINES
September 1952

Calculating Instruments and Machines. Douglas R. Hartree. University of Illinois Press, 1949.

Electronic Analog Computers. Granino A. and Theresa M. Korn. McGraw-Hill Book Company, Inc., 1952.

AN IMITATION OF LIFE
May 1950

Cybernetics. Norbert Wiener. John Wiley, 1948.

MAN VIEWED AS A MACHINE
April 1955

"Solvable and Unsolvable Problems." A. M. Turing in *Science News,* No. 31, pages 7–23. Penguin Books, 1954.